STUDENT GUIDE

EDEXCEL

Geography

Geographical skills
Fieldwork
Synoptic skills

Cameron Dunn and David Redfern

Hodder Education, an Hachette UK company, Blenheim Court, George Street, Banbury, Oxfordshire OX16 5BH

Orders

Bookpoint Ltd, 130 Park Drive, Milton Park, Abingdon, Oxfordshire OX14 4SB

tel: 01235 827720

fax: 01235 400401

e-mail: education@bookpoint.co.uk

Lines are open 9.00 a.m.–5.00 p.m., Monday to Saturday, with a 24-hour message answering service. You can also order through the Hodder Education website: www.hoddereducation.co.uk

ISBN 978-1-4718-6407-0

First printed 2018

Impression number 5 4 3 2 1

Year 2022 2021 2020 2019 2018

Cover photo: Kevin Eaves/Fotolia

Photographs on pp. 94 and 101: David Redfern

Typeset by Aptara Inc., India

Printed in Slovenia

Contents

Getting the most from this book 4
About this book 5

Content Guidance

Geographical skills 6
Fieldwork 49
Synoptic skills 82

Questions & Answers

Assessment overview 93
Questions 94
Glacial landscapes and change 94
Coastal landscapes and change 98
Regenerating places 101
Diverse places 104
Knowledge check answers 108
Index 109

Getting the most from this book

Exam tips

Advice on key points in the text to help you learn and recall content, avoid pitfalls, and polish your exam technique in order to boost your grade.

Knowledge check

Rapid-fire questions throughout the Content Guidance section to check your understanding.

Knowledge check answers

1 Turn to the back of the book for the Knowledge check answers.

Summaries

- Each core topic is rounded off by a bullet-list summary for quick-check reference of what you need to know.

Exam-style questions

Commentary on the questions

Tips on what you need to do to gain full marks, indicated by the icon ⓔ

Sample student answers

Practise the questions, then look at the student answers that follow.

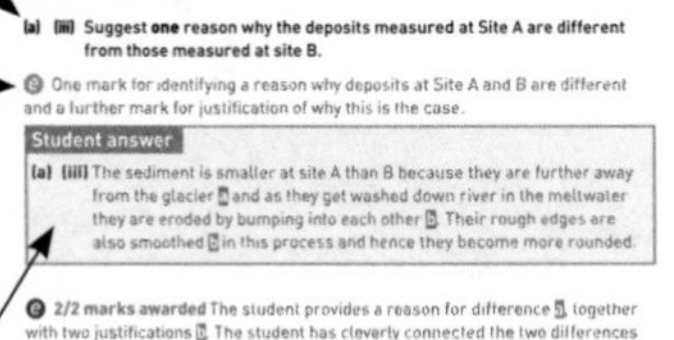

Glacial landscapes and change

ⓔ **2/2 marks awarded** The student identifies the correct hypothesis, and gives the right reason for this.

(a) (iii) Suggest one reason why the deposits measured at Site A are different from those measured at site B. [2 marks]

ⓔ One mark for identifying a reason why deposits at Site A and B are different and a further mark for justification of why this is the case.

Student answer

(a) (iii) The sediment is smaller at site A than B because they are further away from the glacier and as they get washed down river in the meltwater they are eroded by bumping into each other. Their rough edges are also smoothed in this process and hence they become more rounded.

ⓔ **2/2 marks awarded** The student provides a reason for difference, together with two justifications. The student has cleverly connected the two differences in the sediment.

(b) The student collected the data during a one-day field trip in summer. Explain how the student could have gone about collecting the data he needed. [4 marks]

ⓔ One mark for each statement of data collection methodology, as long as the methodology is appropriate in the context of the fieldwork activity.

Student answer

(b) Looking at the photograph, this seems quite a remote area, and so before he set out to collect the data the student should have informed someone of what he was doing and where he was going. Once there, it is clear that some form of sampling should be undertaken. I would suggest using a quadrat — a square metal-framed object which is usually 1 m². The quadrat can be placed in a number of locations across a transect on the deposits — at both sites A and B — possibly two or three sites for each transect. Using either random or systematic sampling within the quadrat, individual items of sediment can then be examined, their size measured and their roundness identified using the Cailleux scale.

ⓔ **4/4 marks awarded** The student makes four valid statements.

(c) You have also carried out field research to investigate glacial landscapes and change. Evaluate the methodologies you used to present and analyse the data you collected. [9 marks]

Location of geographical investigation:

Geographical skills: Fieldwork, Synoptic skills 97

Commentary on sample student answers

Read the comments (preceded by the icon ⓔ) showing how many marks each answer would be awarded in the exam and exactly where marks are gained or lost.

About this book

All students of AS and A-level geography following the Edexcel specification are assessed on the use and application of a range of geographical skills and techniques, fieldwork and broader synoptic thinking. All of these assessments will be undertaken in the context of the geographical concepts that have been studied, and within a variety of content areas, physical and human. These concepts and content areas should also provide inspiration for the group and individual fieldwork activities that have to be undertaken during the course.

Geographical skills and techniques can be assessed in a variety of ways:

(a) in any of the content-based examination questions on any of the AS or A-level papers

(b) in the specific fieldwork and skills questions within the AS examinations

(c) in the Non-Examination Assessment (NEA) that each A-level student has to submit

Fieldwork is assessed in either:

(a) the specific fieldwork and skills questions within the AS examinations

(b) the NEA that each A-level student has to submit

Synoptic skills are assessed within the A-level Paper 3 examination and both AS examination papers.

This guide has two main sections:

Content Guidance — covering the following three areas:

1. Geographical skills: this summarises the specification requirements, and examines each of the required skills and techniques, providing exemplars of most.
2. Fieldwork: this examines fieldwork at AS and A-level. In each case the specification requirements are given, together with advice on how to meet them. At AS, fieldwork and geographical skills are assessed together by means of a written examination. At A-level, fieldwork is assessed by the NEA. Guidance is provided on meeting the requirements of the Edexcel mark scheme for the NEA. This is followed by a number of ideas for individual NEA titles.
3. Synoptic skills: this section looks at the requirements of Paper 3, which examines synoptic themes, and offers advice on answering questions.

Questions & Answers — this includes some sample questions similar in style to those you might expect in the AS exam. There are sample student responses to these questions, as well as detailed analysis, which will give further guidance on what exam markers are looking for to award top marks.

The best way to use this book is to be fully aware of what form of assessment(s) you are doing — i.e. whether you are doing AS or A-level. Then you can select the most appropriate aspects of the book.

Content Guidance

Geographical skills

A range of geographical skills and techniques are tested in a number of ways on the Edexcel specification. They could be assessed in:

(a) examination questions on any of the AS and A-level papers

(b) the Non-Examination Assessment (NEA) — the independent fieldwork task

First we will look at what skills could be assessed — the specification requirements. Then we will look in more detail at the application of each of the **specific skills** identified.

Specification requirements

Competence in using geographical skills should be developed during your study of compulsory and optional topics, not as a separate theme or topic. While the relative balance of quantitative and qualitative methods and skills will differ between topics, you must be introduced to a roughly equal balance of quantitative and qualitative methods and skills across the specification as a whole. This specification requires you to use your prior knowledge and understanding of the geographical, mathematical and statistical skills required at GCSE.

In addition, all of the skills listed below are compulsory and may be assessed across any of the areas of study.

This specification requires you to demonstrate all of the following skills.

1 **Qualitative data skills**
   - **(a)** use and understand a mixture of methodological approaches, including interviews
   - **(b)** interpret and evaluate a range of source material including textual and visual sources, such as oral accounts, newspapers, creative media, social media, aerial, oblique, ground photographs, sketches and drawings
   - **(c)** understand the opportunities and limitations of qualitative techniques, such as coding and sampling, and appreciate how they actively create particular geographical representations
   - **(d)** understand the ethical and socio-political implications of collecting, studying and representing geographical data about human communities

2 **Quantitative data skills**
   - **(a)** understand what makes data geographical and the geospatial technologies (e.g. geographical information systems (GIS)) that are used to collect, analyse and present geographical data
   - **(b)** demonstrate an ability to collect and use digital, geo-located data, and to understand a range of approaches to the use and analysis of such data

(c) use, interpret and analyse geographical information including dot maps, kite diagrams, linear and logarithmic scales, dispersion diagrams, satellite images, GIS

(d) understand the purposes of and difference between the following and be able to use them in appropriate contexts:

  (i) descriptive statistics of central tendency and dispersion, including the Gini coefficient and Lorenz curve

  (ii) descriptive measures of difference and association from the following statistical tests: *t*-tests, Spearman's rank, Chi-squared; inferential statistics and the foundations of relational statistics, including measures of correlation and lines of best fit on a scatter plot

  (iii) measurement, measurement errors and sampling

Exam tip

Many of these skills should be addressed during classroom activities and in your fieldwork. Check back through your notes.

Exam tip

Ensure you have made good use of ICT in finding data and processing them, both in class and at home.

Qualitative data skills

Methodological approaches including interviews

Questionnaire and interview techniques

Carrying out a questionnaire or conducting an interview can yield information. There is an important difference between a questionnaire and an interview, and it is a mistake to use the two terms interchangeably. Questionnaires involve asking a respondent a series of questions, some open, some closed. They typically require written responses although they may be administered face to face. Interviews are one-to-one conversations with a respondent, asking a series of questions that are usually open-ended. In an interview the researcher may ask follow-up questions to explore particular areas of interest that arise. Both techniques are used to collect information about attitudes, perceptions and behaviour that is not available elsewhere as a secondary resource. Questionnaires typically yield some quantitative but mostly qualitative data, whereas interviews are more likely to generate qualitative information.

How you conduct an interview depends on:

- what you wish to investigate: you may need to investigate the root or causal factors of an issue, together with conflicting views and possible solutions
- who you should interview to obtain relevant information or data: for example stakeholders, decision makers or a cross-section of the public. The choice of who to interview is critical

Different types of questions can be used in interviews:

- **hypothetical:** for example, what would you do to resolve the issue/problem?
- **provocative:** for example, do you think that decision-makers are determined to get their way?
- **idealist:** for example, in your opinion what would be the best way to address the issue?
- **interpretive:** for example, what do you mean when you say this is the best approach?
- **leading:** for example, do you think that solution A is better than solution B?

None of these question types is perfect and you should be aware of their deficiencies when analysing your responses. In all cases, however, results are likely to be more reliable if the interviews are based on a stratified sample, whereby you sample the different groups living within a given place (see sampling on p. 12). Such groups may be defined by age, gender, ethnicity and whether they are a visitor or resident (as length of residence may be an important factor).

Source material including textual and visual sources

Oral accounts

An oral account consists of an interview that has been undertaken in the past, and is either in printed form or recorded. It often captures the views and memories of individuals (frequently the elderly) in a community that may have been impacted by change, for example the decline of an industry or change in the make-up of the population of an area, say, through immigration. For these reasons they will be useful when undertaking studies within the themes of Regenerating places or Diverse places. As the speaker is likely to be talking without structure, some form of coding will be needed to abstract useful information from the account (see p. 12).

Many national and local museums, records offices and libraries have built up considerable and well-catalogued collections of what is termed 'oral history'. Since these archives are accessible to the public, you might consider making use of them rather than going out into the field and recording your own.

Exam tip

One useful national source of oral accounts is the British Library, at http://sounds.bl.uk/oral-history

As with other forms of qualitative data, oral accounts should be used with caution. The time lapse between the past and the present can impact on accuracy and reliability of the information. The human memory can be very selective and its quality generally diminishes with age. Many people tend to see the past through rose-tinted glasses, while others have a tendency to remember the more negative aspects of the past.

Newspapers

Newspapers are a useful starting point for the examination of a range of global, national and local events and issues. However, when it comes to the examination of contentious issues then their bias, political or otherwise, must be considered. A number of national newspapers are closely aligned with the 'right' of politics — the *Daily Telegraph*, *The Sun*, the *Daily Mail* and the *Daily Express* — whereas others are more 'left' or 'liberal' – the *Daily Mirror* and the *Guardian*.

Local newspapers may also have some bias, though not necessarily political. They may take sides in the reporting and discussion of local issues, such as whether to use brownfield or greenfield sites for the building of affordable housing, or the worth of proposed schemes for the regeneration of a town centre. The values, perceptions and business interests of the newspaper proprietor or the individual reporter can also be sources of bias.

Creative media

Geographers have often pointed to the importance of the creative imagination in the ways that we respond to places. They have made claims of the power of novels, poems, songs, the visual arts, television, film and video in 'bringing alive' different places.

You may have read a novel set in a place that has also been the subject of a case study in a textbook. The novel is fictional whereas the case study is factual, so it is worth reflecting on the different kinds of knowledge and understanding of that place that you have taken from each of them.

You are encouraged to be open to your interpretation of texts, visual imagery, music and other cultural phenomena. They can indicate the ways in which local cultures shape your ideas about the relationships between humans and the natural world in that place — indeed they provide a form of representation of that place. Furthermore, researchers in these disciplines are also exploring the significance of geography in their work — it is a two-way process.

With these points in mind, you should consider:

- **paintings:** such as those of Constable for rural areas, or Hogarth or L.S. Lowry for urban areas
- **literature:** compare the nature of the rural landscape as described by Jane Austen or Thomas Hardy with the Victorian urban landscapes described by Charles Dickens
- **old postcards:** especially those of the past which can be compared with the same landscape (rural or urban) today
- **television and film:** the great majority of television 'soaps' and dramas rely heavily on their location, which may be imaginary or manufactured, for example *Eastenders*, *Coronation Street*, *Game of Thrones*; or real, such as *Morse* (Oxford), *The Wire* (Baltimore) and *The Sopranos* (New Jersey). The film industry also relies heavily on location — just consider the degree to which places (exotic or otherwise) are a key feature of all the *Bond* and *Bourne* franchise movies.

Exam tip

Choose one form of creative media and reflect on how it portrays one area that you have studied. Was it a fair representation of that place?

Social media

Social media (such as Facebook, Instagram and Twitter) allow people to create, share and exchange information, ideas, opinions and images in virtual communities. Many local places and organisations have online forums where viewpoints as well as practical issues are discussed. Social media might provide some help with your local place investigations by:

- informing about, and perhaps recommending, useful sources of information
- supplying relevant resources such as photos, video clips, newspaper cuttings and commentary
- informing you of the perceptions and lived experiences of different people
- testing views on the success or otherwise of regeneration and rebranding projects

Attachment to place and issues such as regeneration are centred on the perceptions of people (both local residents (insiders) and visitors (outsiders)). Social media allow more views to be sampled than is possible through newspapers, TV and radio programmes. Blogs, online petitions and comments on online news reports can give a more local and 'rounded' assessment of the sense of loyalty to a place. Remember, though, that all material, information and comment from social media come with a 'health' warning. Can you be sure of the authenticity and validity of what you read or are told?

Exam tip

Using the # (hashtag) to accumulate information can be easy, but you are likely to get a lot of qualitative information of variable quality. Select and process it carefully.

As with other such information, it is important that you try to structure your collection of comments so that you are sampling the total population in terms of its components, for example by age, ethnicity and socio-economic groups. You need to portray that a diverse population inevitably has diverse views about the place in which its members happen to live.

Photographs

- **Aerial photographs** are taken from above an area, looking down. Some aerial photographs have been taken with the camera pointing vertically down. Such photographs are very similar to large-scale maps, with a constant scale over the whole area shown. Their main weakness is that they present a level surface — relief features are difficult to see.
- **Satellite images** are one form of aerial photograph that is taken from space. There are several agencies that provide free images, such as NASA and NOAA. Satellite images can show a variety of information and almost all are colour-enhanced. Some may look as if they are photographs showing land uses through a cloud-free sky. Others, however, can show other things such as gas concentrations, thickness of ice, algal growth in water, etc., and so they should be used with careful reference to the colour key that is provided.
- An **oblique aerial photograph** is taken when the camera points at an angle to the ground. The scale on such photographs is more variable. The foreground shows a smaller distance, whereas the areas further away at the top of the photograph have a greater 'real' distance between them. Heights and shapes of features and buildings are more easily seen.
- **Ground photographs** are taken from the ground surface and show a landscape view of an area or place. There are several apps that allow you to annotate such photographs electronically, though of course you can use an overlay to do so physically. Table 1 shows some other ways in which ground photographs can be used.

Table 1 Types of photographs that can be used in a fieldwork context

Photo type	Comment
Black and white	Gives more emphasis to contours in a landscape. Emphasises contrasts more effectively
Photo montage	A series of thumbnail photos that can be used to show a selection of people interviewed, or places visited
360° panorama	Used to show cityscapes or landscapes with a large vista
Photo sequence	Used to show a dynamic event or a series of events over time
Embedding location data	GPS data are embedded into the photograph. Several phone or tablet apps are available for this

Photographs can be obtained from a number of online sources — Getty images, Alamy, Flickr, Facebook and many more. Many local museums and libraries hold collections of old postcards and photographs of street scenes and buildings taken over 100 years ago. Perhaps you will be able to find some for the specific places you are investigating.

Exam tip

One useful national source of past images is Francis Frith at www.francisfrith.com/uk/. It has a collection based in thousands of towns and villages in the UK.

Sketches and drawings

In an examination and for your fieldwork you may be asked to draw a sketch of a landscape (physical or human) or a sketch map. Clearly, it is important that the sketches represent the 'target' with some accuracy, but you are not expected to be a professional artist or cartographer. A key element is the degree to which they are labelled or annotated. A label is a single word, or short set of words, that simply identifies features shown. At AS/A-level you are more likely to be asked to annotate your sketch/sketch map. An annotation requires a higher level of labelling which may be a detailed description, or offer some explanation or even some commentary. Remember to add a title, which should include the location.

Figure 1 shows a field sketch of Cwm Idwal in Snowdonia. It illustrates the difference between a label and an annotation: 'Circular tarn' and 'Arête' are labels; 'Hummocky land with bare outcrops of rock' and 'Evidence of mass movement — slumping' are annotations.

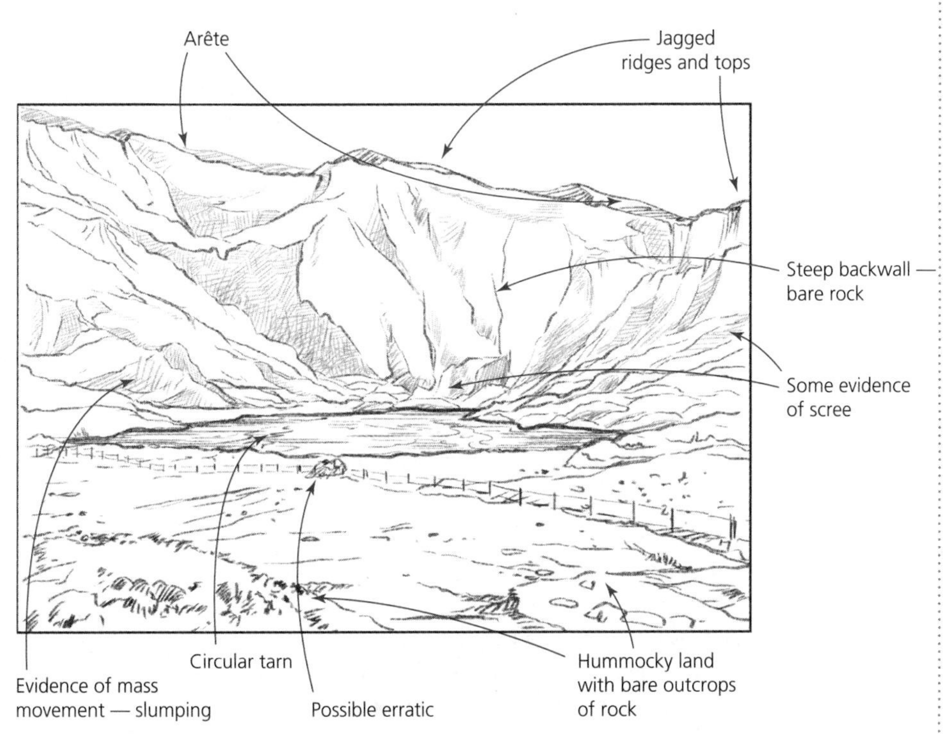

Figure 1 A field sketch of Cwm Idwal, Snowdonia

> **Exam tip**
>
> One simple technique when drawing a field sketch is to divide your paper into four quadrants, and complete the key elements of each quadrant separately. You can rub out the quadrant lines afterwards. Another strategy is to draw in the skyline/horizon first, and then add features coming towards you.

Sketch maps

These are useful for showing the location of a case study or an area of fieldwork (Figure 2). You do not need to include every detail — decide on your priorities. Every sketch map should include:

- title
- scale
- north arrow
- labels/annotations to indicate important features

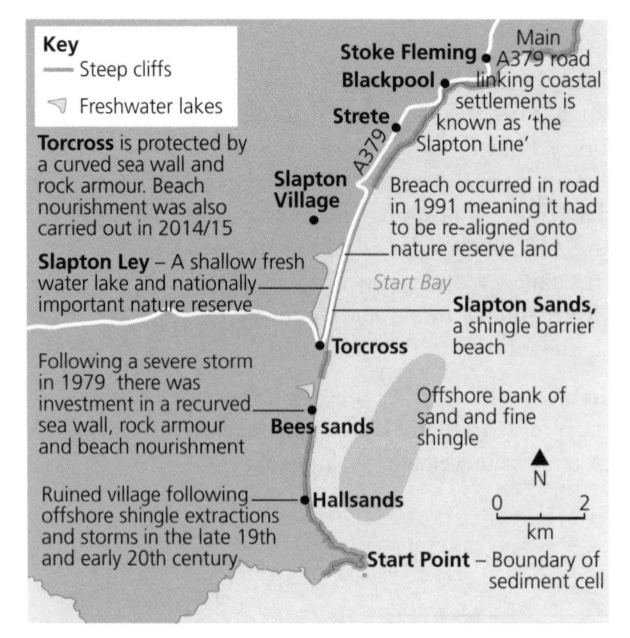

Figure 2 An example of a sketch map — Start Bay, Devon

Exam tip

If you are using a physical overlay on a photograph, map or sketch, make sure you do not write on it while it is covering the image/diagram so that you do not damage the latter.

Coding and sampling

Coding

Coding is the process of organising and sorting qualitative data. It is sometimes referred to as a 'soft' analysis technique as it does not involve mathematical calculations or statistics. Codes serve as a way to label and compile information, which then also allows you to summarise and synthesise what is happening in the data. By linking data collection to the interpretation of the data, coding becomes the basis for developing its analysis.

Coding can be done in any number of ways, but it usually involves assigning a word, phrase, number or symbol (or even using a set of different-coloured highlighter pens) to each coding category. You could go through all your textual data (interview transcripts, direct notes, field observations, etc.) in a systematic way. The facts, ideas, concepts and themes are coded to fit the categories that you have previously identified. Coding can be extremely complicated, but one example could be to review the transcript of an interview, say of a survivor of a natural disaster, and use the codes: Facts, Opinions, Causes, Impacts (Social, Economic, Environmental) and Responses.

Exam tip

Although it is a straightforward task to use colours/highlighters, do not overcomplicate the task by having too many categories.

Sampling

Prior to any investigation, you should make yourself aware of various forms of sampling. The most common ones are:

- random sampling
- systematic sampling
- stratified sampling

Of course, any investigation should consider using a range of sampling strategies, depending on the circumstances of the investigation.

Random means due to or of chance — therefore no pattern should be detectable in any situation. It is used when the environment or population is expected to be similar everywhere. A random sample is one that shows no bias and in which every member of the population has an equal chance of being used or interviewed. Random samples are usually obtained by using random number tables and used to identify items in a list or coordinates on a map.

Systematic sampling is a method in which the sample is taken in a regular way, i.e. every tenth house, every fifth person, or at grid intersections on a map for an area-based sampling exercise. It is often used along transects.

In **stratified sampling**, samples are selected according to some known background characteristic in the statistical population. For example, when studying the distribution of land use types in relation to geology, a stratified sample would select points in proportion to the area covered by each type of geology. If 30% of the area was clay, 25% sandstone, 30% chalk and 15% alluvium then, for a total sample of 200 points, 60 would be selected on clay, 50 on sandstone, 60 on chalk and 30 on alluvium. In this way, the sample is stratified according to a known factor and is thought to be more representative. In investigating the opinions of the local population on a regeneration scheme, the sample should reflect the various interest groups fairly. It could, for example, account for the gender and age distribution in the population, details of which can be obtained from the census.

You may need to make further decisions about sampling. Many investigations rely on a representative sample from the parent population. This population may, for example, be pebbles on a beach, trees in a forest or residents in an area. All samples should be proportional to the size of the total population and so sample size is an important consideration. **Sample size** refers to the number of observations or data points that make up a survey or data set. Very small sample sizes will not reflect the statistical population closely, and so are unreliable and can lead to incorrect interpretations and explanations. Large samples can become unwieldy and difficult to process.

Exam tip

Deciding on the sample size is a crucial decision for all investigations. You must balance validity with practicality.

As stated above, sampling may be random, systematic, stratified or a combination of these. It is important to be able to justify the decisions you make about sampling. Which method of sampling you choose to use depends upon the nature of your investigation. The impression often given is that random sampling is usually best, since it should remove the risk of bias. This is not always the case in geographical investigations, since you will often be looking to recognise some sort of theoretical spatial distribution. This might therefore suggest systematic or stratified sampling as being more appropriate. In studying the downstream changes in a river's discharge there are advantages in having a systematic sample (equal spacing along the river) since you might want to demonstrate that downstream changes take place successively. In applying a questionnaire linked to people's opinions about an issue, you may need to give out questionnaires in proportion to the potential numbers of people in each interest group (stratified sampling), so you do not get skewed results on analysis.

Knowledge check 1

Provide advantages and disadvantages of each of these types of sampling: systematic, random and stratified.

Exam tip

You may be asked to justify your sampling methodology. You should say why you chose one method, and why you rejected other methods.

Remember you can sometimes use more than one sampling technique at a time. For example, if you want to have some structure in your sample but are concerned about potential bias, you might then decide to use some form of random sampling grafted on to a systematic or stratified approach.

Ethical and socio-political considerations

The ethical dimensions of any form of investigation or field research that involves the collection, analysis and representation of geographical information are important. You should think about ethical issues that can arise when you work within communities and in natural landscapes. At all times you should be sensitive to, and show consideration for, the human and physical environment, including the other people found there. A simple way to address this is to always follow the country code when in outdoor areas (such as avoiding trampling and litter), and always be polite to people and respect their views, especially if they say they do not want to be involved in your work. The safeguarding and confidentiality of personal information and data are very important. If working on a human topic, be aware of any social and/or cultural dimensions that might impact on your work. When asking questions, always tell people that you are doing research and why you are doing it, and that their views will be expressed anonymously. Gain their consent before proceeding.

Be careful about the underlying direction, or possible bias, of your questionnaire and interview questions — avoid steering the respondent in one direction. It is often better to ask questions where the outcome is 'in favour' or 'against', 'stay' or 'change' than asking a question which has a view in it, and the respondent has to say 'yes' or 'no' to that view. Some researchers say that an alternative involving 'yes' tends to be skewed in that direction as people prefer to say 'yes' than 'no'. If using secondary data from an available source, consider gaining permission to use the material, if possible.

Exam tip

Construct a table that indicates what you could have done unethically in your investigation(s), and what you did to ensure that it did not happen.

In summary, the ethical dimension of fieldwork can be stated as: do not leave a 'footprint' of where you have been working, either physically or emotionally. Remember you will need to consider the ethical impact of fieldwork in order to fulfil the requirements of the Edexcel mark scheme for the Non-Examination Assessment (NEA).

Quantitative data skills

Geo-located and geospatial technologies including GIS

A global positioning system (GPS) provides location information to a receiver anywhere on or near the Earth where there is an unobstructed line of sight to four or more GPS satellites in space. The system provides critical capabilities to military, civil and commercial users around the world. GIS are underpinned by GPS data, especially when providing real-time data such as vehicle traffic movement. Sat navs in vehicles make use of these data. You may want to use GPS to accurately locate yourself when conducting fieldwork. Most smartphones and tablets have access to such information. Indeed, you can add GPS data to any photographs you may take.

A GIS allows us to visualise, retrieve, question, display, analyse and interpret spatial data to understand relationships, patterns and trends. GIS is essential in the twenty-first century in understanding what is happening and what will happen in geographic space. GIS is used extensively by planning departments (national and local), the police, utility companies and data research companies. There are several free sources of GIS information on the internet, such as Google Earth and Consumer Data Research Centre (CDRC) maps. CDRC maps provide geospatial data from the 2011 census — simply type in the postcode for the area you want to investigate.

GIS is geospatial, meaning it shows layers of data on one map, helping to analyse and understand distributions, patterns and relationships. Maps can be used to compare places or to compare the same place over different time periods. Any information with a locational tag can be used, such as latitude and longitude, addresses and postcodes. Socio-economic and environmental data may be captured, for example population, income, health, education, crime and voting patterns, which may be overlaid on maps with information on satellite images of street layout and terrain.

Exam tip

Place studies involving GIS may need a postcode. One source for this is www.zoopla.co.uk/postcode-finder.

One example of the use of GIS is to investigate crime in an area. Questions you could ask include:

- Where exactly are crimes committed?
- Does the type of crime vary from place to place?
- Are patterns of crime changing?
- Where do criminals live?

The following UK websites provide crime statistics for specific areas and also allow comparison:

- www.crime-statistics.co.uk/: data from 2010; crimes are displayed within a 1-mile radius of the selected postcode
- www.police.uk/: allows a comparison of the crime levels in different neighbourhoods
- http://maps.met.police.uk/: allows users to see what offences (criminal and anti-social) have been reported in local streets

Cartographical techniques

Ordnance Survey (OS) maps

To be able to read an OS map correctly you need to practise a variety of skills:

- grid references (4- and 6-figure)
- scale and estimation of area
- compass direction
- height and relief (contour patterns)
- cross-sections and long sections — both of these are types of simple line graphs and are used to show changes in the shape of the land either across a valley or down a valley. The horizontal (x) axis shows distance and the vertical (y) scale represents height. You should present your section with a realistic scale and make note of the vertical exaggeration if one is present.

Exam tip

A range of scales of OS maps should be used during your course.

It is not necessary to learn all the symbols on an OS map because in an exam you will be supplied with the key. However, it is quicker and easier if you can learn some of them so you won't need to constantly refer to the key.

Synoptic charts

These are weather maps for an area at any one point in time and are likely to be used during your work on the water cycle. You may also consider using one when completing your fieldwork. The key to the symbols on a synoptic chart will always be given in an examination, but you should be aware of the meaning of high and low pressure, and weather fronts (warm and cold). Figure 3 shows an example of a synoptic chart for the UK for a day in winter. There is an area of high pressure over north-west Scotland.

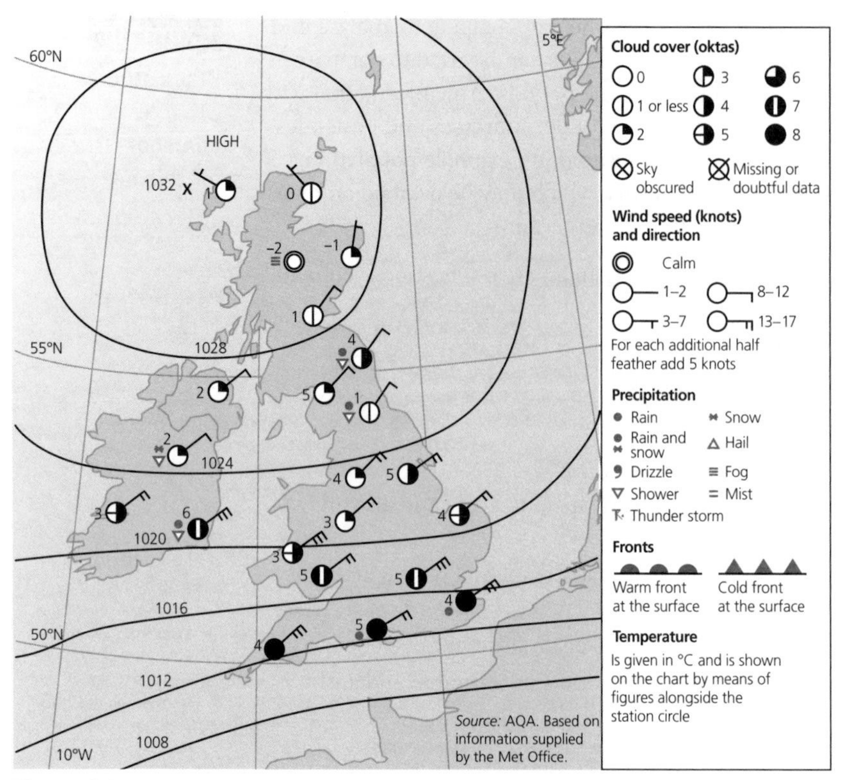

Figure 3 Synoptic chart for the UK on a winter's day

Exam tip

Do not confuse 'synoptic' in a weather context with the word 'synopticity'. They mean totally different things.

Maps with located proportional symbols

Maps can help you to investigate spatial patterns and compare data between different locations. By using a map with proportional symbols it is possible to investigate spatial patterns together with aspects of volume or size of data. Proportional symbols are symbols that are proportionate in area or volume to the value of data they represent. They can take a variety of forms:

- squares
- circles (see Figure 4)
- pie charts
- bar graphs

To make drawing the map worthwhile you need at least three different locations. However, if you have too many it will make the exercise very time-consuming. Try to choose a scale such that overlap is avoided, or minimised. Accurate location is important. Other examples of where these type of maps are useful include looking at downstream changes in discharge along the long profile of a river, or pebble size variations along a stretch of coastline, or characteristics of cities/countries/regions around the world.

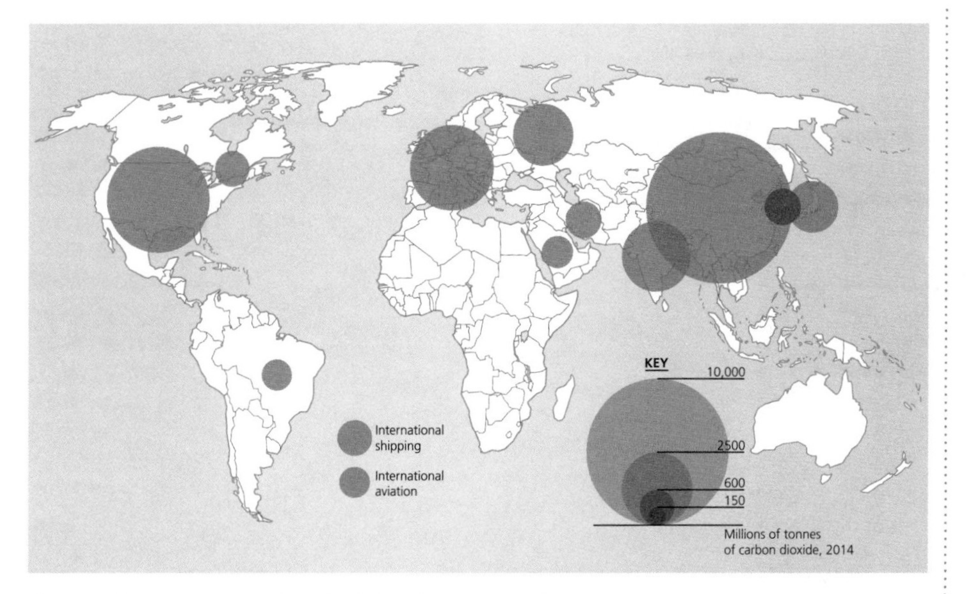

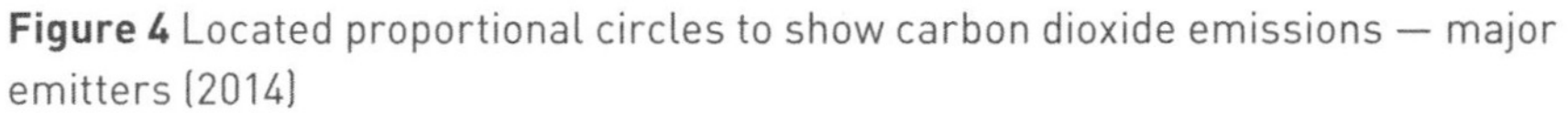

Figure 4 Located proportional circles to show carbon dioxide emissions — major emitters (2014)

Maps showing movement

Movement can be shown on maps in the form of wide arrows, bars or lines. They show direction of movement as well as volume of movement. They are particularly useful in human geography studies.

Flow lines and **desire lines** both show the volume of movement and, in both cases, the width of the line is proportional to the quantity of movement. The difference between them is that a flow line shows the quantity of movement along an actual route whereas a desire line is drawn from the point of origin to the actual destination and takes no account of the actual route.

Flow line and desire line maps may be useful to show:

- migration routes
- movement of traffic across a city (Figure 5)
- tourist destinations
- origins of visitors, workers or shoppers
- flows of money from one place to another (Figure 6)

Trip lines show regular trips or journeys that individual people make. For example, a map can show the places to which people commute from a village location. This technique can be used to work out the catchment areas of shops, schools and other services (see Figure 7).

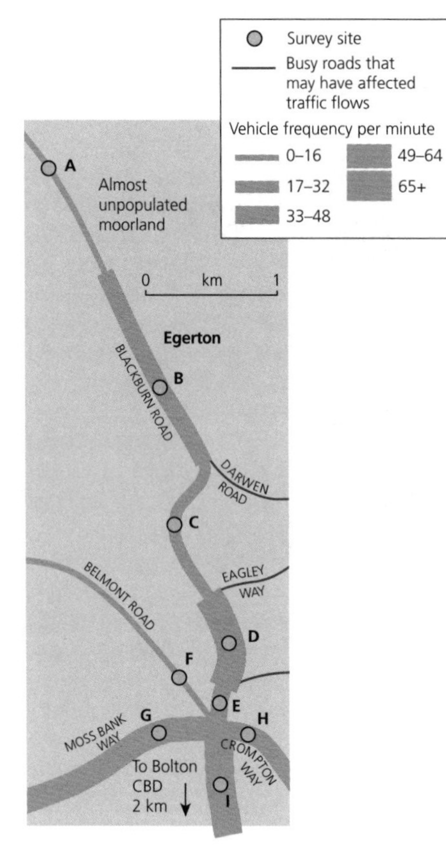

Figure 5 A flow line map showing traffic flows near Bolton

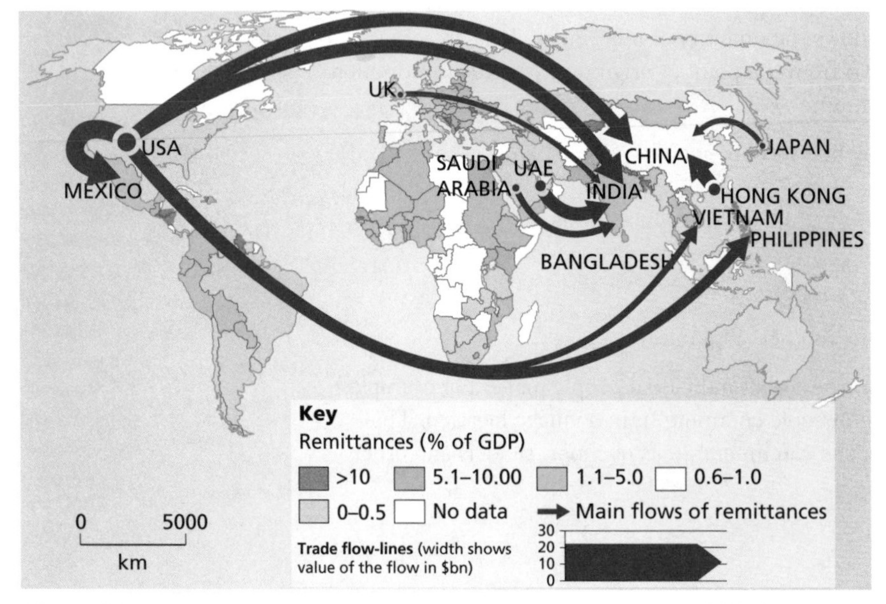

Figure 6 A desire line map showing flows of migrant remittances (2011)

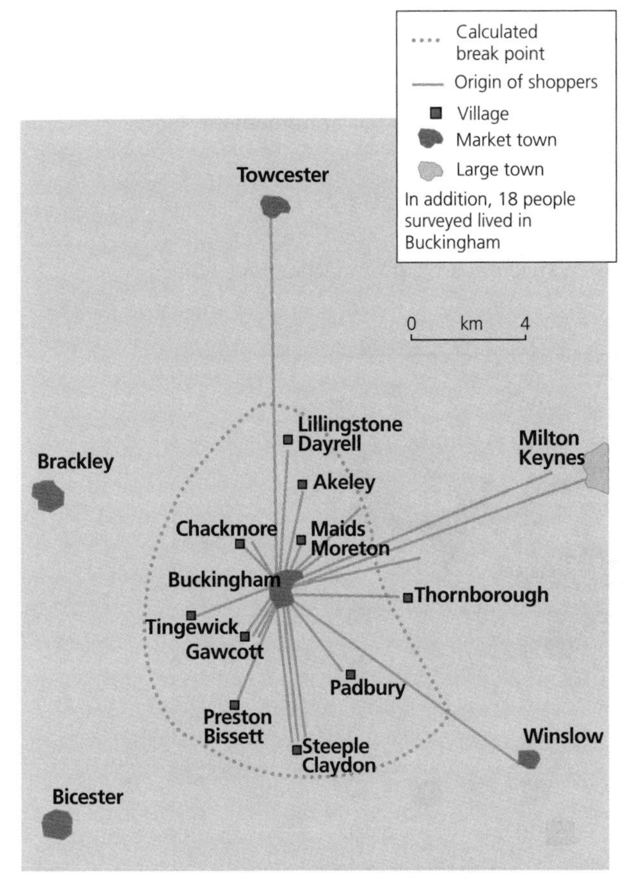

Figure 7 A trip line map showing the origins of shoppers on a market day in Buckingham

Maps showing spatial patterns

A **choropleth map** shows the relative density of a characteristic in an area. Shading using colour, greyscale or line density shows how the data values change from location to location. Choropleth maps are visually striking ways of representing data, as patterns are clearly visible. However, there are also limitations of the technique:

- data are placed in categories or classes and there may be a large variation within the category/class
- it can take time to construct
- if you have too few intervals then a large number of locations may have the same shading, making it difficult to see a pattern
- if you have too many intervals it may be difficult to find enough different shades and again it may be difficult to see a pattern
- it assumes the whole area under one class of shading has a uniform density. In other words, it doesn't show variations that may occur within an area

You may be able to identify the intervals or categories easily. However, in practice, it is usually more difficult. Make sure that you do not include the same number twice, e.g. do not use 1–100 and then 100–200. It should be 0–99.9 and then 100–199.9,

Exam tip

Do not get these three types of map confused. Flow lines show the routes of movements; desire lines the source and destination only; trip lines show movements by individuals.

and so on. One way of determining the interval is by completing a dispersion diagram (see pp. 30–31). It is then easy to see natural breaks where the boundaries of the intervals can be placed. Remember not to choose too many or too few categories — ideally, five to six is best (see Figure 8). Note also that Figure 6 has an underlying choropleth map of percent remittances of national GDP.

Shading can take many forms. If you are using colour or greyscale then convention is to shade from dark to light to represent highest to lowest values. It is better not to use the extremes of black and white — black often suggests a maximum value and white is often used to represent 'no data'.

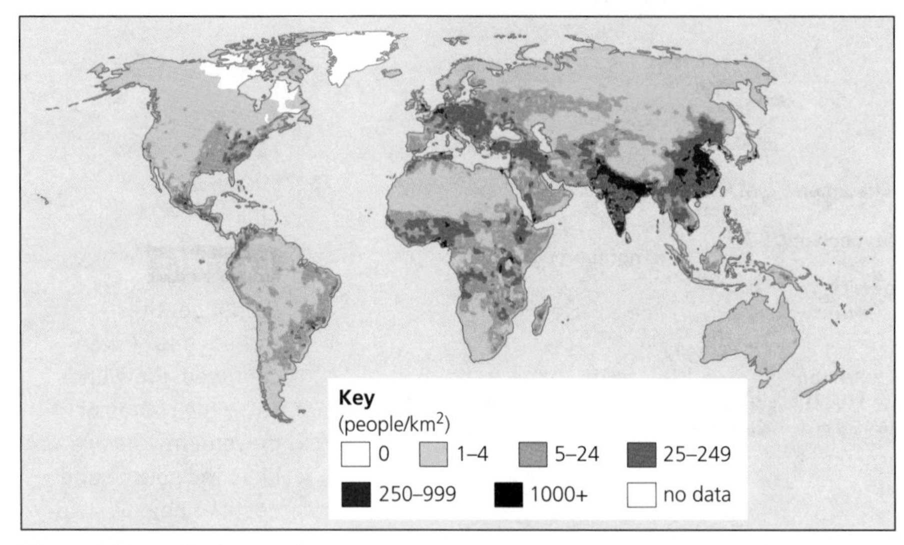

Figure 8 Choropleth map showing global population density estimates, 2015

Exam tip

When describing patterns on choropleth maps, identify areas with high values and areas with low values and then any anomalous situations within each of these areas.

Isoline maps show lines that are drawn on a base map to represent points of equal value. There are many different types and uses of isoline maps:

- contour lines on Ordnance Survey maps
- pressure lines on a synoptic chart (these are called isobars) (see Figure 3)
- temperature (isotherms), for example to show an urban heat island effect
- travel times for commuters (isochrones)
- variations in water depth in a river (see Figure 9)

Isolines are useful for looking at patterns of distribution. However, a large amount of values are needed so they are more suited to group data collection. The more data points you have the more accurate your map will be, although it will be more complicated to draw.

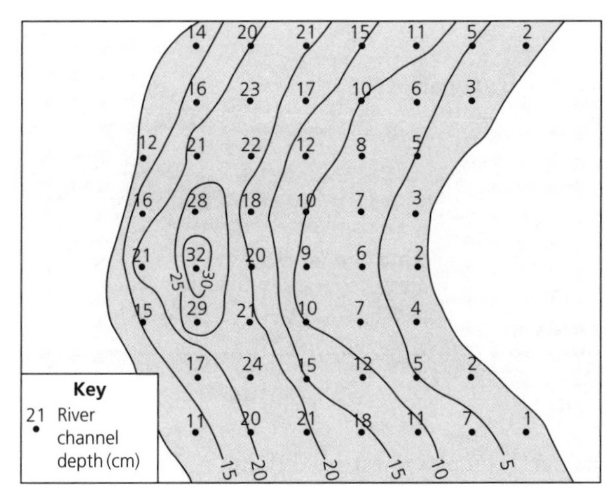

Figure 9 Water depths in a river meander

A **dot map** shows the spatial distribution or density of a variable across an area. For the latter, each dot represents the same value and therefore, unlike choropleth maps, it is possible to estimate the numbers in a particular area by counting the dots. A particular advantage of this type of map is that it gives a clear visual impact of patterns in spatial data and especially the locations of any clusters.

Some examples of where dot maps are useful:

- population or city distribution (Figure 10)
- distribution of ethnic groups
- incidence of disease
- crime rates

Dot maps are easy to draw but they do have some limitations:

- where density is very high it is difficult to count the dots, making a calculation of actual values very difficult
- scale is often an issue — some areas may have densities well below the dot value so will appear empty, appearing to have a value of 0. For this reason, dot maps can be misleading

This type of distribution map can be created more easily today with a GIS system, but hand-drawn maps can still be effective.

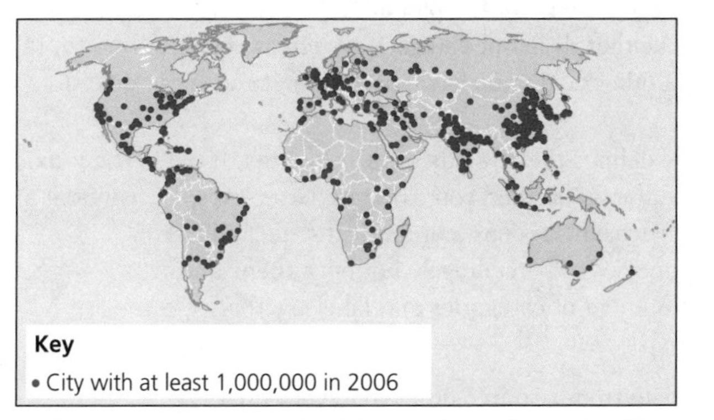

Figure 10 Dot map showing the location of cities of over 1 million people (2006)

Knowledge check 2

What type of map(s) would you use for the following scenarios:

- the population distribution of a country
- the distribution of hospitals in a country
- the variations in ethnicity in a city
- the origins of customers to a large supermarket
- the number of health professionals per 1,000 people around the world
- travel times from a major city such as London
- the main directions of migrations around the world

Graphical techniques

Raw numbers and tables of data (especially large and complex ones) are difficult to understand and interpret. Graphs can tell a story using visual techniques rather than using numbers, and can help the reader understand the meaning in the data. Also, graphs can highlight patterns and trends in data, which not only helps with understanding the data but also allows you to spot anomalies or irregularities. They can help you interpret information at a glance.

General principles

In these days of computer graphics, graphs can easily be produced using spreadsheet software, but you need to be careful to select the most appropriate type of graph to display your particular data. The jazziest type of graph available may not be the best.

Exam tip

It is safest to avoid computer-generated three-dimensional graphs as they can be hard to read and may hide some variables behind each other.

Good practice with graphs (Figure 11) includes the following.

- Axes must be labelled and the graph must have a title.
- The graph area defines the boundary of all the elements related to the graph, including the plot itself and any headings and explanatory text. It emphasises that these elements need to be considered together and that they are separate from the surrounding text.
- The **x-axis** is the horizontal line that defines the base of the plot area. Depending on the type of graph, the *x*-axis represents either different categories (such as years or countries) or different positions along a numerical scale (such as temperature or income).
- The **y-axis** is the vertical line that usually defines the left side of the plot area. If more than one variable is being plotted on the graph then you can have two *y*-axes, one on the left and one on the right. The *y*-axis always has a numerical scale and is used to show values such as counts, frequencies or percentages. For both the *y*- and *x*-axes it is important to choose the right number of categories and labels so that the plot is uncluttered.

The **x-axis** is the horizontal axis on a graph.

The **y-axis** is the vertical axis on a graph.

If the graph you are presenting is based on data from a source other than your own data, then you should acknowledge this somewhere within the graph area or title.

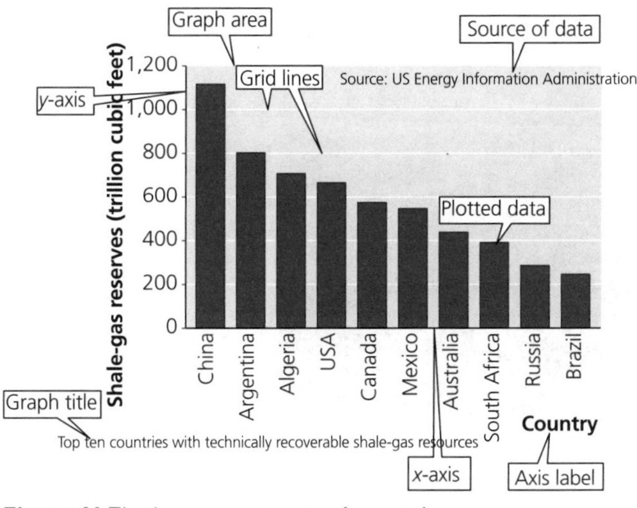

Figure 11 The key components of a graph

Exam tip

As with all forms of data presentation, it is important to be accurate when plotting points on a graph. Inaccuracy will lose (what should be) simple marks.

Line graphs

Line graphs are usually used to show time series data — how one or more variables change over a continuous period of time. Examples include monthly rainfall or annual unemployment rates. Line graphs are good for identifying patterns and trends in the data. They can also be used for displaying continuous spatial data, e.g. how pollution levels vary with increasing distance from a source, or how the number of pedestrians changes with increasing distance from a central business district (CBD).

In a **simple line graph** the *x*-axis represents the continuous variable (for example, year or distance from the initial measurement) whilst the *y*-axis scale shows units of measurement of the factor being recorded. Several data series can be plotted on the same line graph and this is particularly useful for analysing and comparing the trends in different, but comparable, data sets. These are **comparative line graphs** (Figure 12).

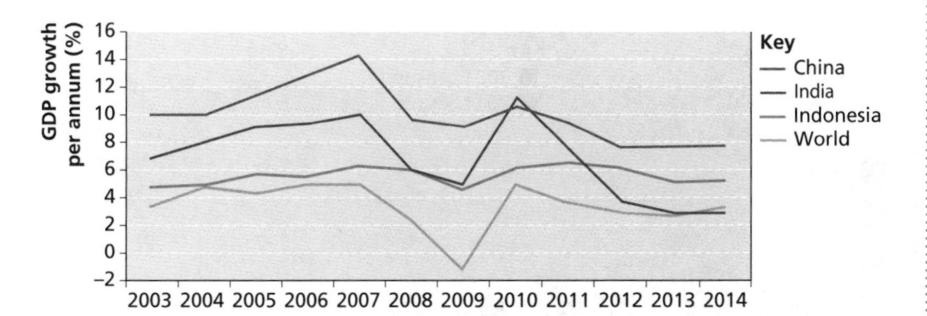

Figure 12 Annual economic growth for China, India and Indonesia compared with world average

Sometimes data can be plotted as a **compound line graph** where categories are placed one on top of the other. Be careful in the interpretation of these types of graph as you need to read the different areas separately, often indicated by different shading or colours (Figure 13).

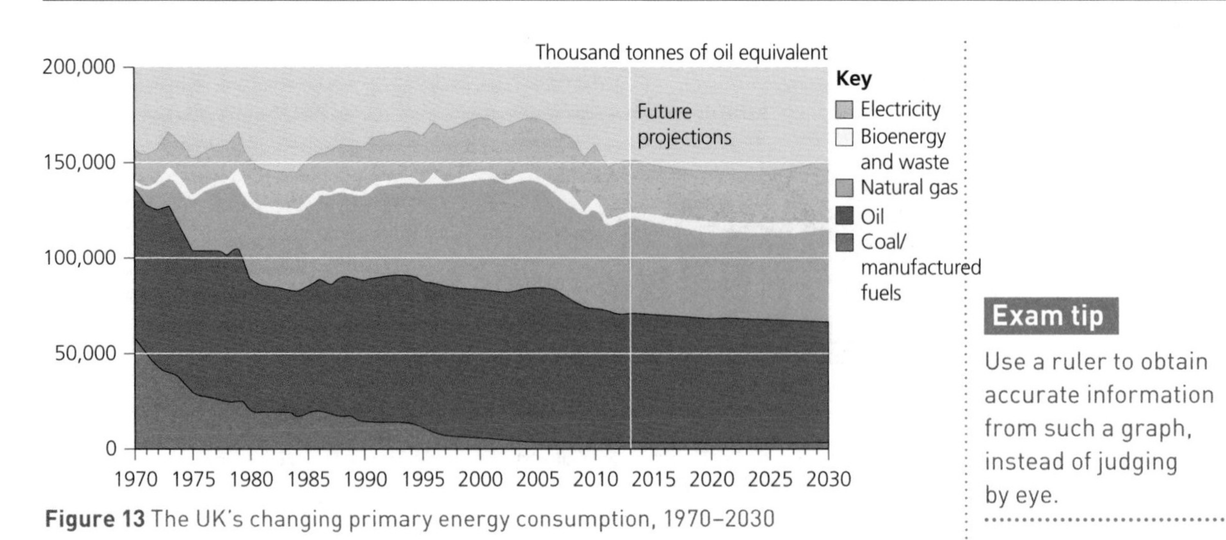

Figure 13 The UK's changing primary energy consumption, 1970–2030

Exam tip

Use a ruler to obtain accurate information from such a graph, instead of judging by eye.

Line graphs can be **divergent**, i.e. they can show positive and negative trends (see the World line on Figure 12).

Bar graphs

A bar graph, in its simple form, is used to show the differences in frequencies or percentages among discrete categories of data. In a bar graph the categories (*x*-axis) are displayed as rectangles or blocks of equal width, with their height proportional to the frequency or percentage of the data.

Bar graphs are useful for comparing categories of a variable within different groups, for example, a comparison of data over two time periods. Such a **comparative** bar graph would have two bars, one for each of the two time periods, and the height of each bar would be scaled (Figure 14). Bar graphs can also be **divergent**, i.e. they can show positive and negative values (Figure 15) and they can be **compound**. A common form is a percentage compound graph (Figure 16).

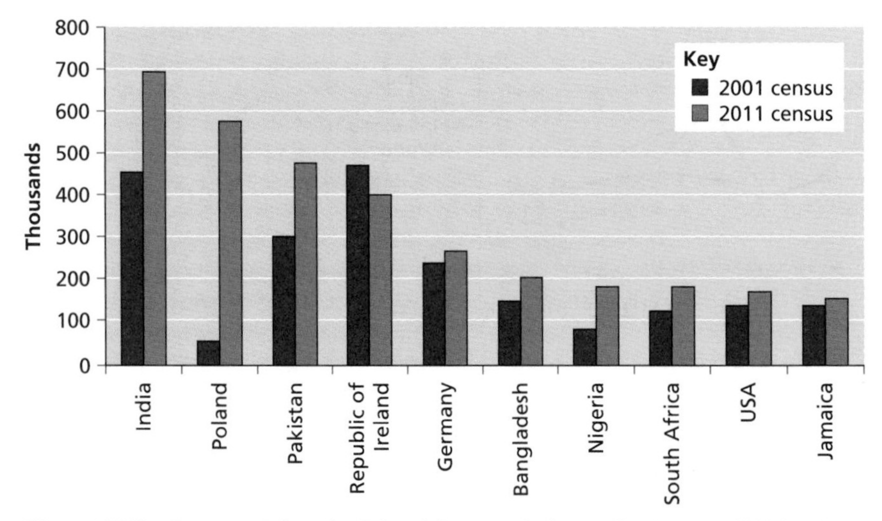

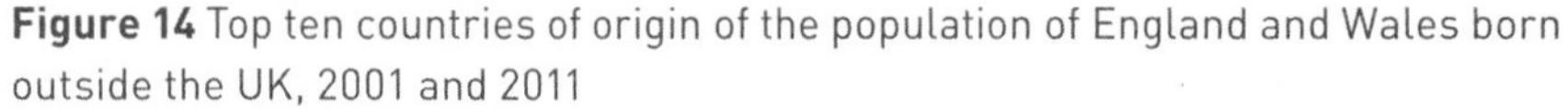

Figure 14 Top ten countries of origin of the population of England and Wales born outside the UK, 2001 and 2011

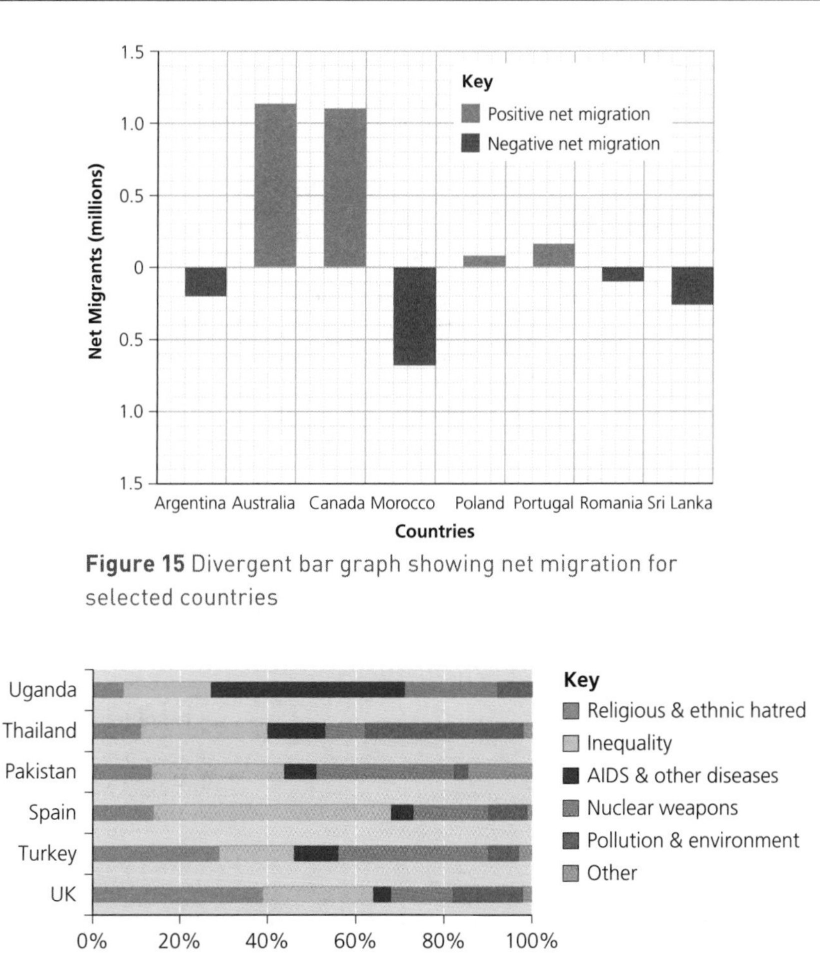

Figure 15 Divergent bar graph showing net migration for selected countries

Figure 16 Pew Research Centre's 2014 Global Attitudes survey: what is the greatest threat to the world?

Scatter graphs

Scatter graphs are used to show the relationship between two sets of measurements or data. For example, a scatter plot might be used to present information about the relationship between income equality and life expectancy (Figure 17). Regression lines or lines of best fit can be added to the scatter graph. These help the user to understand the strength of the relationship between the two sets of data. Note the line of best fit does not have to go through the graph's origin. It can also show either a positive or a negative relationship.

Scatter graphs are often used to establish whether it is worthwhile to carry out further work on the data through hypothesis testing and statistical analysis. You should present the **independent variable** (the quantity which you think controls the relationship) on the x-axis and the **dependent variable** (the quantity which varies in response to the independent variable) on the y-axis. So, if you thought that plant growth was controlled by the amount of sunlight, you would plot sunlight hours on the x-axis and plant growth rates on the y-axis.

A **dependent variable** is data that are affected by the change in the other variable.

An **independent variable** is data that are expected to cause the change in the other variable.

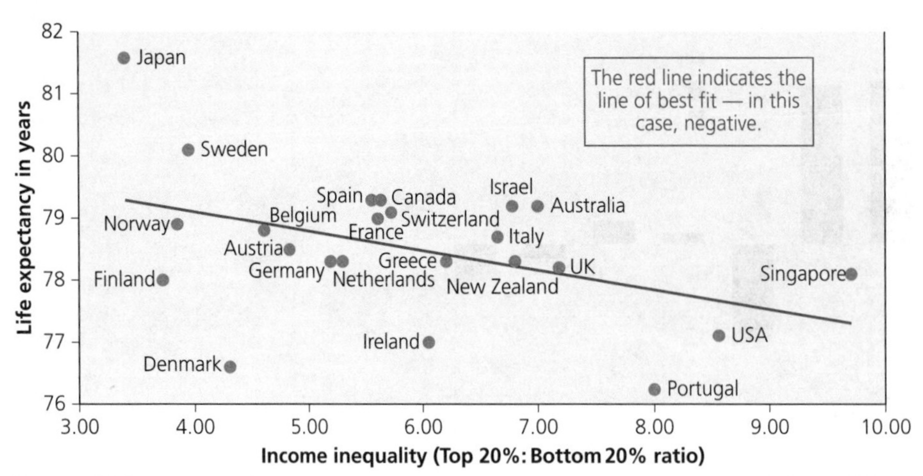

Figure 17 Scatter graph to show the relationship between income inequality and life expectancy

Points lying some distance from the line of best fit are anomalous, and are called **residuals**. For the ease of establishing a possible best-fit relationship, these residuals can be ignored (as long as there are no more than two or three of them).

A **residual** (or an anomaly) is a point/feature that does not fit the general trend or pattern.

Exam tip

Scatter graphs may show a relationship that has no causal link — it is just coincidental. Further research may be needed.

Pie charts

Pie charts are a common technique to show a frequency distribution. In a pie chart, the frequency or percentage is represented both visually and numerically, so a reader can quickly understand the data and what the researcher is conveying (Figure 18). However, it is true that pie charts tend to be the default option for many students, and they are not always the best way to show data. They can become unwieldy, especially when there are too many categories of data. Pie charts with fewer than four categories are often better presented as tables. Use them with caution and keep them simple. The overlaying of individual percentages, for example, can add too much clutter and should be used sparingly.

When a number of pie charts are drawn with their size proportional to the total values being represented, they are called **proportional divided circles**. In this case, the size of the circle is determined by the formula

$$r = \sqrt{\frac{v}{\pi}}$$

where r is the radius of the chart, and V is the total value shown. See Figure 4 on p. 17 for an example.

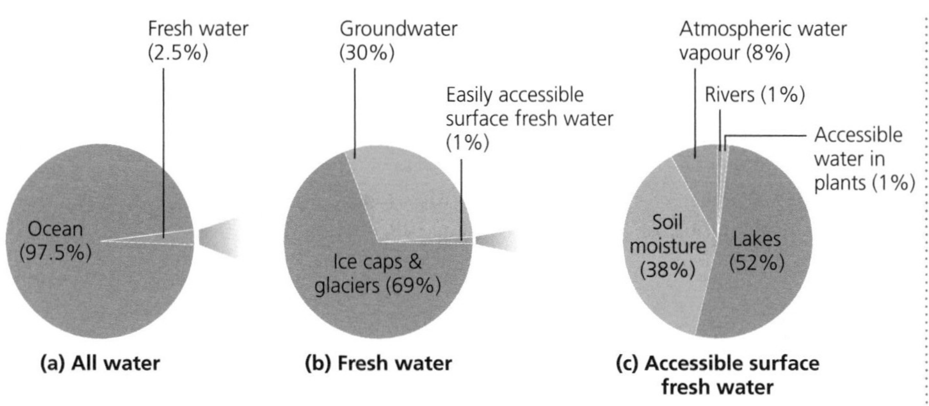

Figure 18 The global 'water pot'

Exam tip

When constructing a series of proportional divided circles, it is good practice to begin the sequence of segments at 12 o'clock, work clockwise and maintain the same sequence of segments to make comparison easier.

Triangular graphs

Triangular graphs, as the name suggests, are graphs that are constructed in the form of an equilateral triangle (Figure 19). They can only be used if the following conditions apply, and for these reasons their use is fairly limited:

- the data must be able to be divided into three component parts
- the data must be in the form of percentages
- the percentages must total 100

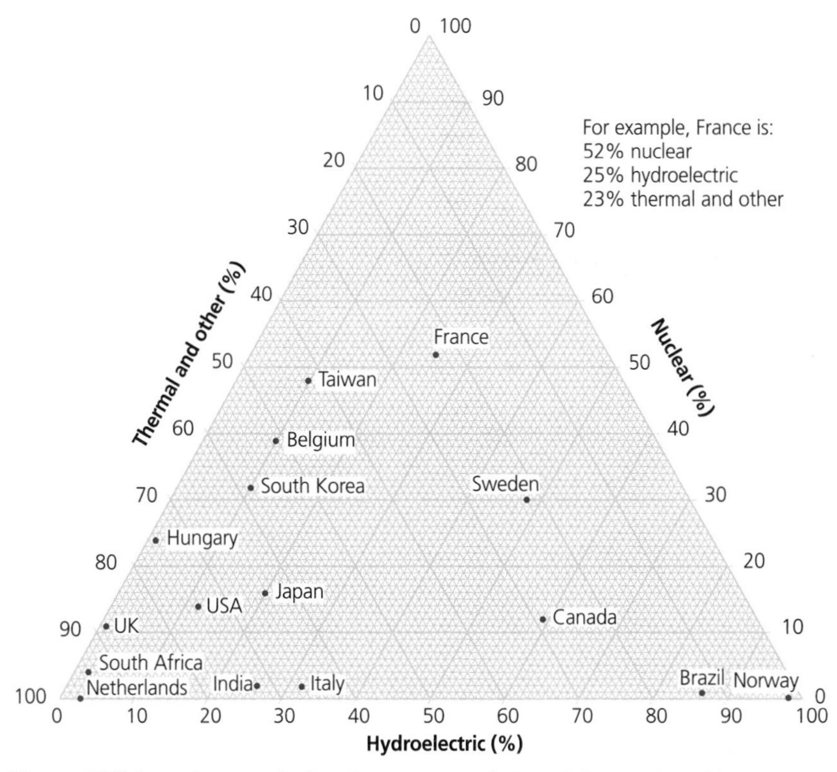

Figure 19 Triangular graph showing percent of electricity produced by generating source for selected countries

Each side of the triangle represents one axis and one component, and measures from 0% to 100%. From each axis, lines are drawn at 60° angles to carry the values. Triangular graphs are useful because you can show a large amount of data on one graph. As with pie charts it is easy to see relative proportions and identify the dominant variable. However, triangular graphs can be difficult to interpret and care must be taken not to make errors reading off the incorrect axis. On the other hand, they have an in-built checking system, as all values for one plot must total 100. If the total is not 100, an error has been made.

Exam tip

One way to read a triangular graph is to imagine them as three separate pyramids, one for each variable. Each side acts as a base for that variable (0%) and each apex represents 100%.

Graphs with logarithmic scales

Logarithmic graphs are a useful form of line graph where a large range of data has to be plotted. In an arithmetic line graph the scale increases by equal amounts; **logarithmic scales** differ from this in that the scales are divided into a number of cycles, each representing a ten-fold increase. An example of this principle is: the first cycle ranges 1–10, the second cycle will be 10–100, the third 100–1,000, and so on. When both the *x*- and *y*-axis scales are logarithmic they are called log-log graphs. If only one axis is logarithmic, then it is known as a semi-logarithmic graph.

A **logarithmic scale** is a scale that is divided into cycles or intervals that increase ten-fold each time.

One disadvantage of these graphs is that you cannot use positive and negative values on the same graph. Figure 20 shows an example of semi-logarithmic graph paper where the logarithmic scale is shown on the vertical *y*-axis. Examples of the use of logarithmic scales include the Moment Magnitude Scale used to measure earthquakes and the Hjulström curve used in river studies.

Exam tip

Logarithmic graphs are useful when interpreting rates of change — the steeper the line the faster the change.

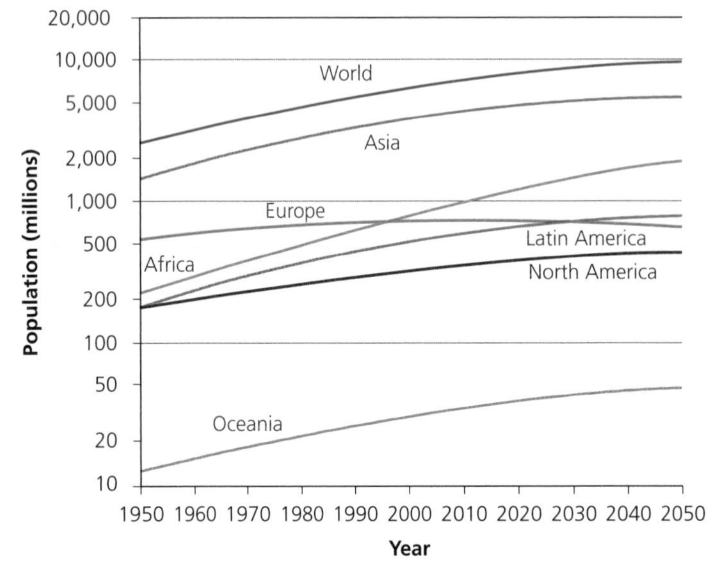

Figure 20 World population growth and projected population growth in different continental regions, 1950–2050

Radial diagrams/charts

These graphs are particularly useful when one variable is a directional feature, such as rose diagrams which show both the direction and frequency of winds. The circumference represents the compass directions and the radius can be scaled to show the percentage of time that winds blow from each direction. In a similar way they can be used to show the orientation of features, such as corries (cirques) in a glaciated area (Figure 21).

Radial diagrams can also be used when one variable is a recurrent feature, such as time period of 24 hours or an annual cycle of activity. Hence they can be used to plot traffic flows over a period of time during a day, or over a year.

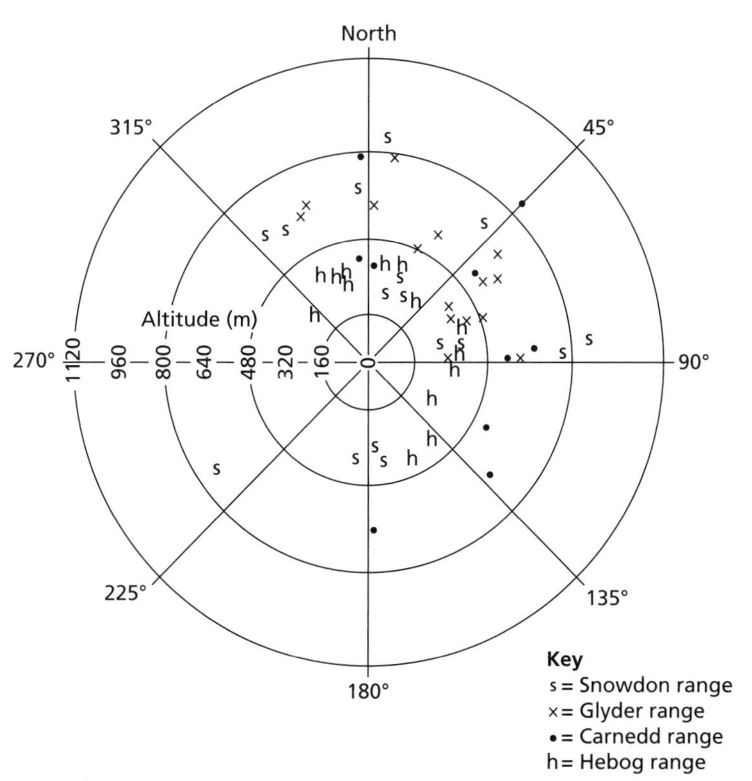

Figure 21 Radial diagram: cirque orientation against altitude, Snowdonia

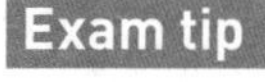

Radial diagrams can only be used where the scale around the circumference is continuous, such as time and direction.

Kite diagrams

Kite diagrams are a useful way of showing changes over distance, particularly vegetation types along a transect. One axis is used for distance and the other for the proportions (usually percentage of cover) of individual plant species (Figure 22). The width of the kite, representing a single species, enables a visual comparison to be made of the distribution of vegetation at any point along the transect.

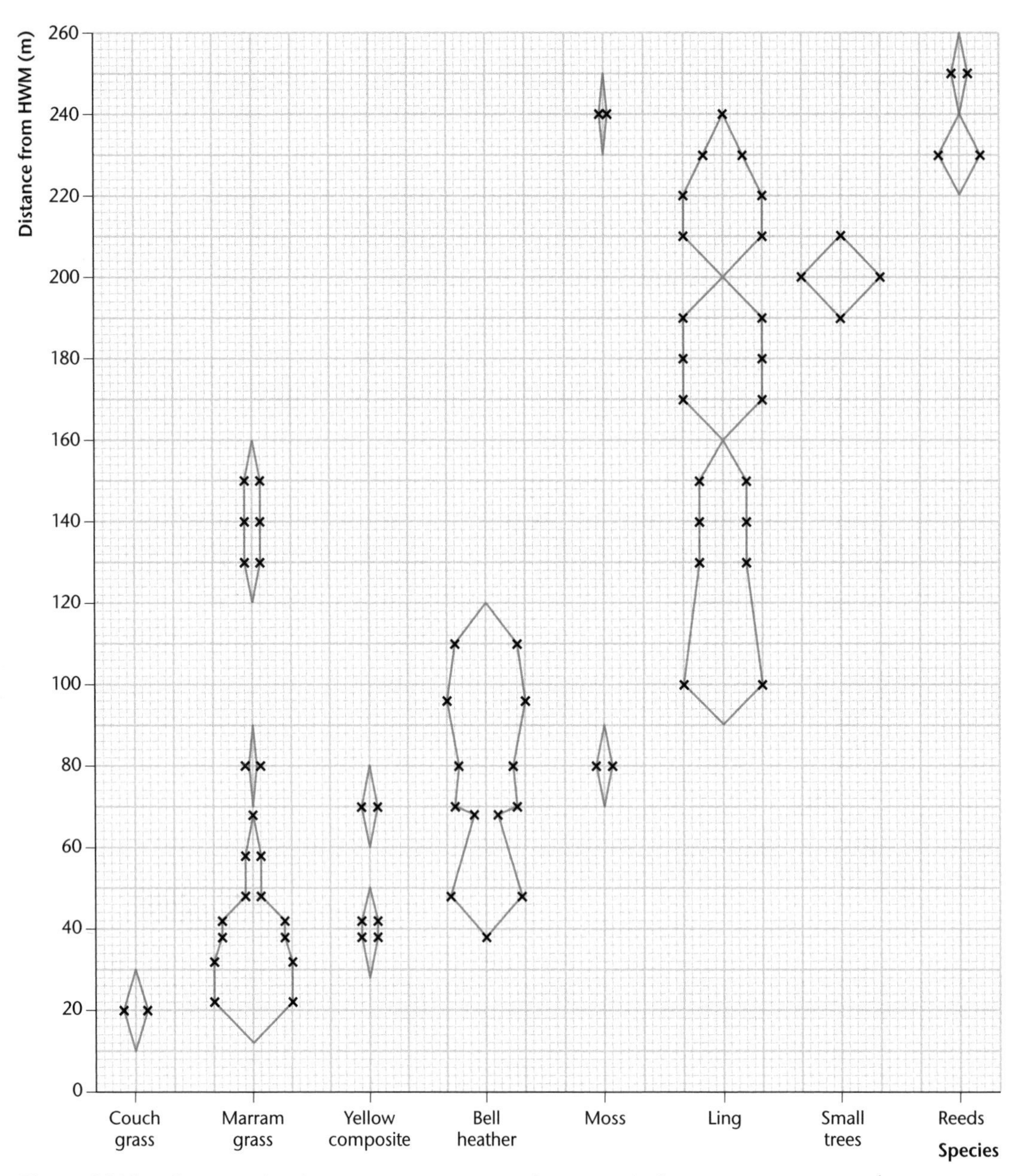

Figure 22 Kite diagram showing a transect across a dune area in Dorset

Dispersion diagrams

Dispersion graphs are useful for comparing sets of data (especially patterns both within and between sets) for two or more locations. There are numerous occasions when dispersion graphs should be used, and they are often under-used by geography students. A key advantage of dispersion graphs is that the data are clearly laid out along a vertical axis as a basis for further analysis. In Figure 23, for example, it is clear that the corries on the Isle of Arran have a wider range of angles of orientation than the corries in the Glyders of Snowdonia (data given in Table 2).

Exam tip

The construction of a kite diagram with more than four plant species needs careful planning.

Table 2 The lip orientation of 15 corries in the Glyders of Snowdonia and 15 on the Isle of Arran (°)

Corrie number	1	2	3	4	5	6	7	8	9	10	11	12	13	14	15
Glyders	30	60	45	55	75	50	80	50	10	15	10	35	45	50	85
Arran	5	5	10	55	15	30	95	5	185	70	120	40	30	115	110

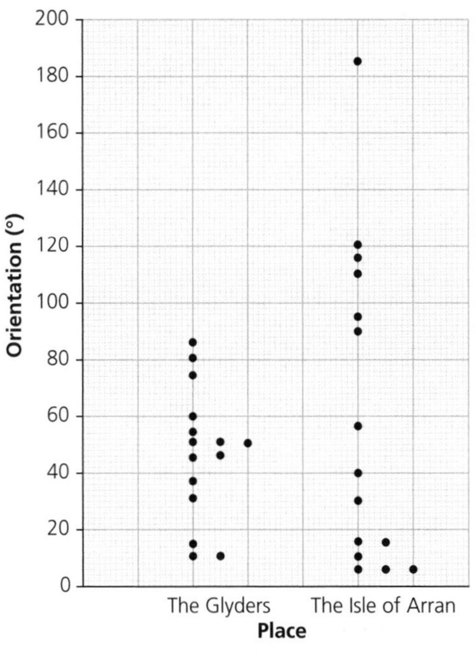

Figure 23 Dispersion diagram of corrie lip orientation for the Glyders (Snowdonia) and the Isle of Arran

Knowledge check 3

What type of graph(s) would you use for the following scenarios?:

- the changing length of a glacier/sand spit/alluvial fan over time
- the percentage of global deaths from a range of non-communicable diseases
- the energy mix of a country
- the top ten countries with the highest urban populations
- the potential relationship between life expectancy and health expenditure of a nation

Statistical techniques

For much of the following section, refer to the data in Table 3.

Table 3 Discharge in four months (1994–2012) for a river in northern England (cumecs)

Year	January	April	July	October
1994	56.68	15.23	14.04	72.61
1995	70.96	6.35	6.15	42.59
1996	86.42	25.85	6.55	72.93
1997	77.86	7.50	1.88	67.91
1998	24.45	45.05	40.87	30.57
1999	69.94	24.68	6.03	45.62
2000	33.26	27.11	31.89	70.23
2001	81.19	17.29	42.80	47.27
2002	34.51	31.16	3.60	42.68
2003	88.81	13.95	18.01	50.07
2004	49.33	34.81	10.48	33.12
2005	31.69	31.58	6.16	27.56
2006	79.70	42.05	21.00	11.98
2007	79.46	50.62	5.16	31.17
2008	93.93	12.78	7.17	25.22
2009	21.16	16.78	9.56	41.18
2010	10.00	10.00	3.60	10.20
2011	29.98	17.58	31.00	80.60
2012	99.70	32.60	32.05	46.65

Measures of central tendency

Measuring central tendency refers to ascertaining a measure of the 'middle' value of a data set. There are three ways of measuring central tendency:

- mean
- mode
- median

These techniques are very useful to geographers, enabling us to summarise a data set by giving the mid-value or most frequently occurring data. They can also be used as part of more complex techniques, such as inter-quartile range and standard deviation.

Mean

Formula:

$$\bar{x} = \frac{\Sigma x}{n}$$

The mean (sometimes called the average) is calculated by totalling all the values in a data set and dividing the sum by the number of values in the set. The mean is particularly useful if the data have a small range. However, if the range is large then the mean will be heavily influenced by the extreme values and could give a distorted picture.

Referring to Table 3, the mean discharges for the four months are:

- January: 58.90 cumecs
- April: 24.37 cumecs
- July: 15.68 cumecs
- October: 44.75 cumecs

Mode

This is the value that occurs most frequently in a set of data. You need to know all values before calculating the mode. Mode is of no use if there are no repeating values. There may be more than one mode — this is called 'bi-modal'. The mode is often useful when classifying data into groups or classes. It is useful to see which classification occurs most frequently. This is called the 'modal class'.

There are no repeating values in Table 3, so no modes can be given.

Median

This is the middle value in a data set. The data need to be placed in rank order before you can calculate the median.

If there is an odd number of values, perform the following calculation to work out the median value:

$$\frac{n+1}{2}$$

where n is number of values in the data set.

For example, if you have 23 values in a data set the median will be the 12th value in the rank order. If the number of values is even, the median is the mean of the middle two values. So, if there are 24 values, add the values for the 12th and 13th positions and divide by 2.

The median value often needs to be supported by other techniques, such as inter-quartile range (IQR). However, unlike the mean, it is not affected by extreme values.

Referring to Table 3, which has 19 sets of data for each month, the median discharges for the four months are the 10th item of data (when ranked highest to lowest):

- January: 69.94 cumecs
- April: 24.68 cumecs
- July: 9.56 cumecs
- October: 42.68 cumecs

None of these measures give an accurate picture of the distribution of data. On their own they are of limited value. However, the above example shows that some limited judgements can still be made. In two of the measures the central tendency for river discharge is largest in January and lowest in July. The spread of data is however quite large for January, and the extreme values of 10.00 and 99.70 may be making the mean value higher. The uneven spread of data for January, as shown in Table 3, means that this data set is 'skewed'. In general, the greater the skew in a data set the greater variation in the three measures of central tendency.

To improve the usefulness of the above calculations, measures of the dispersion or variability of the data should be calculated.

Exam tip

Be clear about the differences between the three indices of central tendency, and of the different ways in which they are calculated.

Measuring dispersion

These techniques are used to measure the spread of data. Range, inter-quartile range and standard deviation allow you to analyse your data in more depth, looking at how spread the data are around either the mean or the median.

Range

This is simply the difference between the highest value and the lowest value. It gives you a basic idea of the spread of data but, like the mean, it is affected by extreme values. An anomaly can therefore give a false picture. Referring to Table 3 the ranges are as follows:

- January: 89.70 cumecs
- April: 44.27 cumecs
- July: 40.92 cumecs
- October: 70.40 cumecs

Therefore we can see that January has the largest range and July the smallest.

Inter-quartile range (IQR)

The inter-quartile range is a measure of dispersion around the median. It is worked out by ranking the data (highest to lowest) and placing the data into quarters, separated by quartiles, on a dispersion diagram (Figure 23). The top 25% of the data is placed above the upper quartile (UQ) and the bottom 25% is placed below the lower quartile (LQ). The inter-quartile range or IQR is the difference between the 25% and 75% quartiles, or UQ minus LQ (Figure 24).

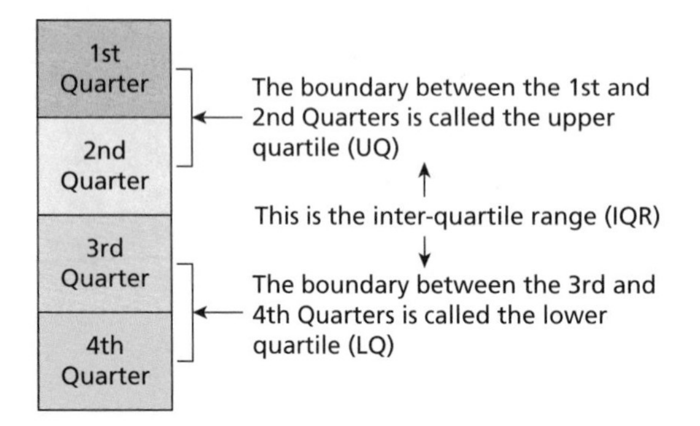

Figure 24 Inter-quartile range

The IQR is more useful than the range in indicating the spread of data as it removes any extreme values (i.e. those occurring in the 1st Quarter and 4th Quarter) and considers the spread of the middle 50% of the data around the median value. There is a series of formulae to calculate the IQR.

Upper quartile (UQ) = $\frac{n+1}{4}$th position in the data set (ranked from highest to lowest)

Lower quartile (LQ) = $\frac{n+1}{4} \times 3$th position in the data set (ranked from highest to lowest)

Inter-quartile range (IQR) = UQ – LQ

Referring back to Table 3:

	January	April	July	October
Upper quartile	81.19	32.60	31	67.91
Lower quartile	31.69	13.95	6.03	30.57
IQR	49.50	18.65	24.97	37.34

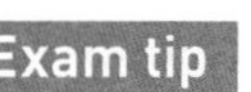

Exam tip

Geography has long had the convention of ranking data from highest to lowest. You will find it more straightforward to do so.

There is least variation in April and most in January.

Standard deviation

Standard deviation (SD) is a measure of the degree of dispersion about the mean.

Formula for standard deviation:

$$\sigma = \sqrt{\frac{\Sigma(x-\bar{x})^2}{n}}$$

where σ is standard deviation, Σ is sum of, $\bar{x}$ is mean and n is number in the sample.

It is possible that two sets of data could have the same mean but have a very different spread of data. SD will tell you the extent of this spread — in other words, how reliable the mean is. A low SD indicates that the data points tend to be very close to the mean; a high SD indicates that the data are spread out over a larger range of values and the mean is therefore less reliable as there is obviously a lot of variation in the sample. Hence SD is used when you want to compare the dispersion of two or more sets of data.

An important aspect of SD is that it links the data set to the **normal distribution** (Figure 25).

Normal distribution is a theoretical frequency distribution that is symmetrical about the mean.

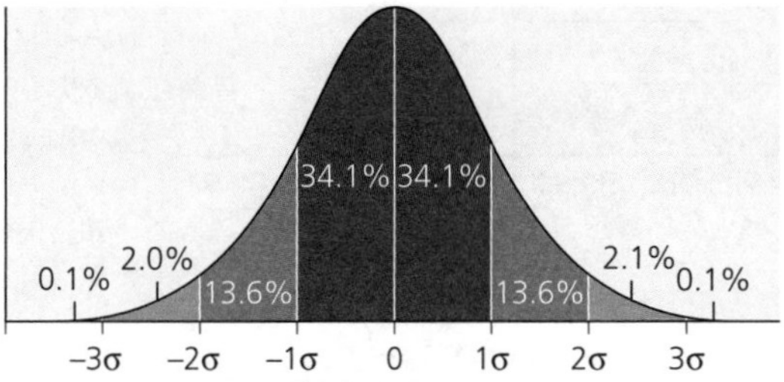

Figure 25 The normal distribution (sometimes known as the bell-shaped curve)

Exam tip

The link between SD and the normal distribution allows you interpret the outcome more easily. Try to remember these percentages.

In a normal distribution:

- 68% of the values lie within ±1 standard deviation of the mean
- 95% of the values lie within ±2 standard deviations of the mean
- 99% of the values lie within ±3 standard deviations of the mean

Worked example of standard deviation

Refer back to our data in Table 3. In Table 4, the standard deviation has been calculated for the river discharge in January.

Table 4 Standard deviation calculation for the river discharge in January

January discharge (cumecs)	$x - \bar{x}$	$(x - \bar{x})^2$	
56.68	-2.22	4.93	
70.96	12.06	145.44	
86.42	27.52	757.35	
77.86	18.96	359.48	
24.45	-34.45	1,186.80	
69.94	11.04	121.88	
33.26	-25.64	657.41	
81.19	22.29	496.84	
34.51	-24.39	594.87	
88.81	29.91	894.61	
49.33	-9.57	91.58	
31.69	-27.21	740.38	
79.90	21	441	
79.46	20.56	422.71	
93.93	35.03	1,227.10	
21.16	-37.74	1,424.31	
10.00	-48.90	2391.21	
29.98	-28.92	836.37	
99.70	40.80	1,664.64	
$\Sigma x = 394$	$\bar{x} = 58.90$	$\Sigma(x - \bar{x})^2 = 14,458.91$	$\frac{\Sigma(x - \bar{x})^2}{n} = \frac{14,458.91}{19}$
			$\sigma = 27.59$

Standard deviation for the river discharge in January = 27.59 cumecs.

The other SD calculations are:

- April: 12.59 cumecs
- July: 13.10 cumecs
- October: 19.96 cumecs

This exercise in SD suggests that there is more clustering around the mean in the spring and summer months. The mean is therefore more reliable at these times. This is also supported by the other measures of central tendency and dispersion, which indicate that there is less dispersion in spring (April) and summer (July) and more dispersion in autumn (October) and winter (January).

The Gini coefficient

One format of the use of dispersion is the Gini coefficient. This is a statistical measure that can be used to assess the extent to which the distribution of income

Exam tip

An Excel spreadsheet will complete the calculation for you. If you use a calculator, do not rush, thereby avoiding errors.

Knowledge check 4

Distinguish between the inter-quartile range (IQR) and standard deviation (SD).

among the people of a country varies from a perfectly equal distribution (income inequality). Values vary between 0 (perfect equality) and 100 (perfect inequality). These extremes are practically impossible, so we must say that the lower its value, the more equally household income is distributed, and conversely the higher the value, the more unequally household income is distributed. The Gini coefficient is therefore a measure of the overall extent to which households, from the bottom of the income distribution upwards, receive less than an equal share of income.

How is it represented and calculated?

It is expressed graphically by the **Lorenz curve** of the household income distribution, from which the Gini coefficient can be calculated. Based on a ranking of households in order of ascending income, the Lorenz curve is a plot of the cumulative share of household income against the cumulative share of households.

A **Lorenz curve** illustrates the degree of unevenness in a geographical distribution.

The curve will lie somewhere between these two extremes:

- **complete equality**, where income is shared equally among all households, results in a Lorenz curve represented by a straight 45° line
- **complete inequality**, where only one household has all the income and the rest have none, is represented by a Lorenz curve that comprises the horizontal axis and the left-hand vertical axis

The Gini coefficient is the ratio of the area between the Lorenz curve of the income distribution and the diagonal line of complete equality, expressed as a proportion of the triangular area between the curves of complete equality and inequality (see below for explanation).

A Lorenz curve

A Lorenz curve is drawn on graph paper and makes use of cumulative percentage data (Figure 26). The vertical axis carries the cumulative data, where points are plotted in the order of the largest first, which is then added to the second largest, then to the third largest, and so on. The horizontal axis simply records the cumulative process. The plots are then connected by a line. If another line is drawn onto the graph to represent an even distribution, then the degree of unevenness can be seen. The greater the deviation of the plotted line from the line of even distribution, the greater the degree of unevenness. A highly concave Lorenz curve represents a high level of unevenness, and therefore high level of concentration.

To explain how a Lorenz curve is drawn, consider a country with ten regions, A to J, each of which contains a different proportion of that country's population (Table 5). How uneven is the population? To find this out, we can draw a Lorenz curve.

Table 5 Percent of population in regions of country X

Region	A	B	C	D	E	F	G	H	I	J
Percent of population in region	10.5	1.6	12.2	1.8	35.3	6.7	2.1	25.3	3.5	1

The first plot on the graph (Figure 26) is shown as E — 35.3% of the population, at the first point on the cumulative process. The second plot is H — at 60.6% (35.3 + 25.3), at the second point. This cumulative process carries on until the final plot is 100% at the tenth point. All of the plots are joined by a line that appears to describe a 'curve'. The line of complete equality is then added, and the shaded area represents the degree to which the unevenness occurs. To calculate the Gini coefficient, the following formula is used:

$$\text{Gini coefficient} = \frac{\text{Area A}}{(\text{Area A} + \text{Area B})}$$

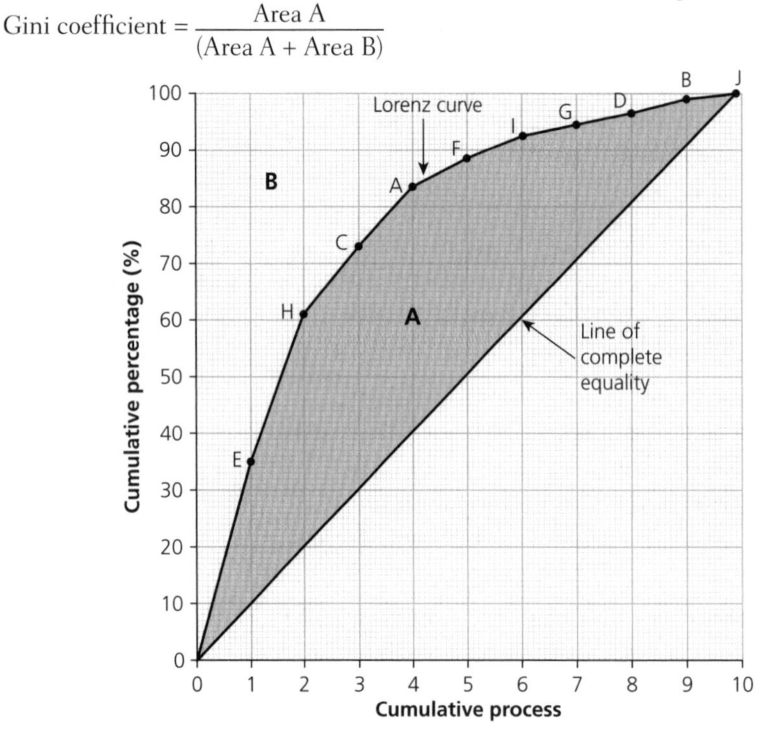

Figure 26 Lorenz curve of percent population distribution in country X (by region)

How is the Gini coefficient used?

It is highly unlikely that you will be asked to calculate a Gini coefficient, although you could be asked to construct a Lorenz curve. It is more likely that you will be provided with a table of Gini coefficients (Table 6), possibly alongside other data, and asked to explain what the data tell you. Here you have to demonstrate your understanding of **number** (see also p. 46).

Table 6 Wealth inequality between and within selected countries (2014)

HDI rank		GDP per capita (nominal, US$)	Share of national income going to *poorest fifth* of population (%)	Share of national income going to *richest tenth* of population (%)	Gini coefficient
High HDI					
12	Sweden	49,000	12	30	25
14	UK	46,000	6	29	38
Medium HDI					
91	China	7,600	4	35	37
108	Indonesia	3,500	8	29	33
135	India	1,600	8	31	34
Low HDI					
180	Uganda	700	4	42	44
164	Burundi	300	7	33	33

Exam tip

Simple mathematical errors can be made in the cumulative process. Double check your addition before plotting any points.

The United Nations regularly produces figures for nations in terms of their income inequality using the Gini coefficient. Their interpretation of such data is:

- low inequality: under 30 (e.g. Sweden and Hungary)
- relatively low inequality: 30–39.9 (e.g. UK and China)
- relatively high inequality: 40–44.9 (e.g. Russia and Uganda)
- high inequality: 45–49.9 (e.g. Venezuela and Mexico)
- very high inequality: 50–59.9 (e.g. Nigeria and Kenya)
- extremely high inequality: 60 and over (all of the countries in this group are in southern Africa)

Inferential and relational statistical techniques

Comparisons are often made between two sets of data to see if there is a relationship between them. However, we should always recognise that the existence of a statistical relationship does not prove a causal link. Further investigation is needed to establish whether or not there is a causal link — the statistical evidence may just suggest there could be.

The Student's t-test

Sometimes in fieldwork it is not possible to collect as much data as you would like because of limitations in numbers or time available. In such cases it is still possible to compare two sets of data using this statistical test.

The Student's *t*-test is a way of comparing two samples to test a hypothesis that there is a difference between two (small) sets of data. It should be used when there are no more than 30 individuals or sites in the sample. Mathematically, the *t*-test uses the difference in the **standard errors** of the two mean values. It assumes a distribution in the data which is symmetrical about a standard error that is zero, as in the normal distribution. However, it is flatter than the normal distribution and becomes progressively flatter with the smaller numbers involved.

By definition all samples must have a degree of inaccuracy built into them. This inaccuracy is quantified by the standard error of the mean. It is usually worked out by calculating the standard deviation of the data and then dividing that by the square root of the number of the sample.

The formula for the Student's *t*-test is:

$$t = \frac{|\bar{x}_1 - \bar{x}_2|}{\sqrt{\frac{s_1^2}{n_1} + \frac{s_2^2}{n_2}}}$$

where: $\bar{x}_1$ and $\bar{x}_2$ are the means of two data sets (note the difference must be absolute — i.e. positive), S_1 and S_2 are the standard deviations of those data sets, n_1 and n_2 are the numbers in the two data sets.

Procedurally you should start by stating your hypothesis, such as: *there is a difference between the two sets of data A and B*. You should then be prepared to accept the opposite — *there is no difference between A and B*. This is called the **null hypothesis**.

Exam tip

The specification does not require you to know about the standard error of the mean, although it does expect that you understand the concept of 'error'.

Standard error is the potential difference between the mean value calculated from a sample of a data set, and the mean value of the total population of the data set.

Exam tip

You will not be expected to learn the formula for Student's *t*-test for an examination.

The **null hypothesis** is usually that 'there is no difference between variable 1 and variable 2'.

You then need to do another test to check whether your result could have occurred by chance — this means 'how **significant** is your result'. To do this you compare your result with a table of critical values (see Table 8). You also need to decide which significance level you are going to use. For geographical purposes, you would usually use the 0.05 significance level. This means that there is a 5-in-100 likelihood of the results occurring by chance. Or, to put it another way, if other researchers completed the same experiment, 95 out of 100 would get the same result — therefore there is a strong chance of a difference between the two items. You then need to see whether your result is **above** the critical value. If your value is below the critical value you must accept the null hypothesis — i.e. you cannot be sure that the difference is significant.

Significance is the degree to which you can be confident that your results did not occur by chance. Usually the 95% (0.05) and 99% (0.01) confidence levels are used.

A further complication is to ascertain the **degrees of freedom**. Where the two sample sizes are A and B, the degrees of freedom are:

A (– 1) + B (– 1)

In statistics, the number of values that are free to vary in the final valuation of a statistic is called **degrees of freedom**.

Exam tip

The specification does not require you to know about degrees of freedom. They are simply used to work out which set of critical values to compare your result with.

A worked example

A vegetation survey was undertaken in an area of north-west England to see if there was a significant difference in the number of plant species supported by acid moorland and limestone upland. Using a quadrat the students counted the number of species found within the frame at ten sites on acid moorland and eight sites on limestone upland. The results are shown in Table 7.

Table 7 Number of plant species on sample sites of acid moorland and limestone moorland

Quadrat	Number of species on acid moorland (x_1)	Number of species on limestone moorland (x_2)
1	6	14
2	8	12
3	9	6
4	4	11
5	7	15
6	11	14
7	7	17
8	6	8
9	8	
10	7	
Σ	73	97
Mean ($\bar{x}$)	7.3	12.13
n	10	8
Standard deviation (σ)	1.79	3.44

Knowledge check 5

Referring to Table 7, calculate the modal values for each of the columns x_1 and x_2.

$$t = \frac{4.83}{1.34}$$

The result is $t = 3.60$

Degrees of freedom = (10 – 1) + (8 – 1) = 16.

Table 8 Significance tables for the Student's *t*-test

Degrees of freedom	Significance level 0.05	Significance level 0.01
1	6.31	63.7
2	2.92	9.93
3	2.35	5.84
4	2.13	4.60
5	2.00	4.03
6	1.94	3.71
7	1.89	3.50
8	1.86	3.36
9	1.83	3.25
10	1.81	3.17
11	1.80	3.11
12	1.78	3.06
13	1.77	3.01
14	1.76	2.98
15	1.75	2.95
16	1.75	2.92
17	1.74	2.90
18	1.73	2.88
19	1.73	2.86
20	1.73	2.85

A result of 3.60 is greater than the **critical value** of t at both the 0.05 significance level and the 0.01 level. A null hypothesis that there is no difference between the two data sets can be rejected. Therefore there is a significant difference in the number of plant species supported by acid moorland and by limestone moorland. The latter has more variety.

The **critical value** is used with many statistical techniques to test the significance (or confidence) level. Each technique has a table of critical values against which the result of the statistical test is compared.

The Spearman's rank correlation test

One way of showing a possible relationship between two sets of data is a scatter graph (Figure 17). Spearman's rank is another way of testing a relationship. When you draw a scattergraph you can see by eye if there is a relationship, but you will probably not be able to clearly assess the strength of the relationship as many points may be some distance from the line of best fit. Spearman's rank correlation is used to test the strength of the relationship between two sets of data, providing you with a numerical value between 0 and +1, or between 0 and –1. Once you have this figure you can then test its significance — the likelihood of your results occurring by chance.

The test can be used with any set of raw data or percentages but it is only suitable if all the following criteria apply:

- two data sets which you believe may or may not be related
- at least ten pairs of data
- no more than 30 pairs (as this makes the exercise unwieldy)

Start by stating your hypothesis, such as: *as X increases then so does Y*. You should then be prepared to accept the opposite — the null hypothesis, in other words (see p. 39).

A worked example is given below. Once you have completed the table and have your result at the end of the calculation you should have a figure between –1 and +1. This indicates the strength and type of relationship:

- a result close to +1 indicates a positive relationship (i.e. as one set of data increases so does the other)
- a result close to –1 indicates a negative relationship (i.e. as one set of data increases the other decreases)
- a result close to 0 means there is no relationship and you must accept the null hypothesis

Exam tip

In an examination you are not expected to learn the formula for Spearman's rank.

Strengths of Spearman's rank:

- it gives an outcome that is objective
- it enables you to demonstrate a numerical relationship between two sets of data, although it must be stressed that it may not be a causal relationship
- you can state whether or not the relationship is significant
- it is less sensitive to anomalies in data as each piece of data is ranked — large differences could only be one rank apart

Weaknesses of Spearman's rank:

- it does not tell you whether there is a causal link (i.e. that change in one variable leads to a change in another), only that a relationship exists
- too many 'tied ranks' can affect the validity of the test
- it could be subject to human error such as inaccurate calculations

Worked example

You are investigating the relationship between the percentage of unskilled workers and the percentage unemployed to ascertain the variations in social and economic conditions in an urban place you are studying. The null hypothesis is that *there is no relationship between the percentage of unskilled workers and the percentage of unemployed in an area*. You construct Table 9.

Table 9 Variations in social and economic conditions in an urban area in England

Ward	A Unskilled workers %	Rank (A)	B Unemployed %	Rank (B)	Difference in rank (A – B) (*d*)	d^2
1	5.5	13	9.4	15.5	2.5	6.25
2	6.6	10	15.8	6	4	16
3	6.0	11	11.7	13	2	4
4	2.2	17	8.9	17	0	0
5	15.6	1	23.0	1	0	0
6	8.4	6	18.9	3	3	9
7	7.3	8	13.9	10	2	4
8	8.5	5	15.1	8	3	9
9	8.9	4	14.2	9	5	25
10	11.6	3	17.5	4	1	1
11	12.5	2	22.0	2	0	0
12	5.6	12	12.4	11	1	1
13	8.1	7	15.4	7	0	0
14	1.8	18	6.2	18	0	0
15	7.2	9	16.7	5	4	16
16	4.3	15	12.3	12	3	9
17	4.2	16	11.6	14	2	4
18	5.1	14	9.4	15.5	1.5	2.25

$\Sigma d^2 = 106.5$

$$\text{Spearman's rank correlation coefficient } (R_s) = 1 - \frac{6\Sigma d^2}{n^3 - n}$$
$$= 0.890$$

(Note that Wards 1 and 18 have tied ranks for % unemployed.)

You now need to do another test to check whether your result could have occurred by chance — this means 'how significant is your result'. To do this you have to compare your result with a table of critical values (Table 10). First look at the number of pairs of data you have — in this case there are 18. You then need to decide which significance level you are going to use. For geographical purposes, you would usually use the 0.05 significance level. This means that there is a 5-in-100 likelihood of the results occurring by chance. Or, to put it another way, if other researchers completed the same experiment, 95 out of 100 would get the same result — therefore there is a strong chance of a relationship between the two items.

You then need to see whether your R_s result is **above** the critical value for the number of pairs you have. If your R_s value is below the critical value you must accept the null hypothesis — i.e. you cannot be sure that your relationship is significant.

Exam tip

As elsewhere, do not rush your work here and thereby avoid errors.

Knowledge check 6

Referring to Table 9, calculate the ranges for each of the columns A and B

Table 10 Critical values for R_s

n	0.05 Significance level	0.01 Significance level
10	±0.564	±0.746
12	0.506	0.712
14	0.456	0.645
16	0.425	0.601
18	0.399	0.564
20	0.377	0.534
22	0.359	0.508
24	0.343	0.485
26	0.329	0.465
28	0.317	0.448
30	0.306	0.432

We can see that in our example the R_s value of 0.890 is well above the 0.05 significance level of +0.399. It is also well above the 0.01 significance level of +0.564. This means that there is a very low (1-in-100) likelihood of the results occurring by chance and we can reject the null hypothesis. Having rejected the null hypothesis, you can accept that there is a strong relationship between the percentage of unskilled workers and the percentage unemployed and that it is highly significant.

Exam tip

Be clear in your understanding of the purpose of significance testing.

The Chi-squared test

The Chi-squared test is used to investigate spatial distributions. It looks at frequencies or the distribution of data that you can put into categories, for example pebble shapes at different sites along a beach or frequencies of plant types at different stages of a vegetation succession. Chi-squared is a comparative test as it compares actual data collected against a theoretical random distribution of the data.

The data collected are called the **observed data** (**O**).

The theoretical, random, distribution is called the **expected data** (**E**).

For a Chi-squared test to be conducted:

- the data need to be organised into categories
- the data cannot be in the form of percentages and must be displayed as frequencies
- the **total** amount of observed data must exceed 20
- the expected data for each category must exceed 4

As with other statistical tests, Chi-squared requires and tests a null hypothesis. The null hypothesis is: *there is no significant difference between the observed distribution and the expected distribution.*

The strength of Chi-squared lies in the fact that, as with other statistical tests, you can check the significance of your results. Also, as with other statistical tests, weaknesses include human error in calculating x^2. It also does not explain why there is, or there is not, a pattern to the distribution. As with Spearman's rank, this will need further investigation.

Chi-squared (x^2) is calculated using the formula:

$$x^2 = \sum \frac{(0 - E)^2}{E}$$

Worked example

A group of students investigated the orientation of pebbles in an exposed bed of glacial till. The glacial till was situated near the lip of a corrie in the Lake District. The students wanted to investigate whether there was a pattern to the orientation of the long-axis of the till. Their hypothesis was: *there is a significant trend in the orientation of pebbles within the glacial till.* They measured the orientation of 40 pebbles and placed their results into four categories:

1 0–45° = 2 pebbles
2 46–90° = 10 pebbles
3 91–135° = 23 pebbles
4 136–180° = 5 pebbles

The data suggest that there is a preferential direction but as this may be due to chance, a Chi-squared test was carried out. The test begins with the null hypothesis: *there is no significant difference between the observed orientation of pebbles and an expected random orientation.* Next the students created Table 11.

Table 11 Working for the Chi-squared test

			A	B	C
Orientation	**Observed (0)**	**Expected (E)**	**0 – E**	**(0 – E)²**	**$\frac{(0-E)^2}{E}$**
0–45°	2	10	–8	64	6.4
46–90°	10	10	0	0	0
91–135°	23	10	13	169	16.9
136–180°	5	10	–5	25	2.5
					x^2 = 25.8

The Chi-squared value $x^2 = 25.8$.

By itself, the result is meaningless. The students needed to test its significance. To do this, they worked out the degrees of freedom using the formula $(n - 1)$, where n is the number of observations; in this case the number of categories which contained observed data. For this example $n = 4$, so the degrees of freedom are $(4 - 1) = 3$.

Exam tip

The specification does not require you to know about 'degrees of freedom'. They are simply used to work out which set of critical values to compare your result with.

Using Table 12 showing the critical values for Chi-squared, the students compared their result for x^2 at the degree of freedom of 3 for the 0.05 and 0.01 significance levels. If the x^2 result is **the same as or greater than** the value given in the table, then the null hypothesis can be rejected.

Exam tip

You will not be expected to learn the formula for Chi-squared for an examination.

Exam tip

Be clear in your understanding of the purpose of significance testing.

Table 12 Critical values of Chi-squared

Degrees of freedom	Significance level	
	0.05	0.01
1	3.84	6.64
2	5.99	9.21
3	7.82	11.34
4	9.49	13.28
5	11.07	15.09
6	12.59	16.81
7	14.07	18.48
8	15.51	20.09
9	16.92	21.67
10	18.31	23.21
11	19.68	24.72
12	21.03	26.22
13	22.36	27.69
14	23.68	29.14
15	25.00	30.58

The students wrote a summary statement to express the result for the Chi-squared test:

At 3 degrees of freedom, the x^2 result of 25.8 is above the 0.01 critical value of 11.34. Therefore we can reject the null hypothesis and accept that the orientation of the till did not occur by chance and is not randomly orientated. There is therefore a significant trend in the orientation of the pebbles.

Knowledge check 7

What type of statistical technique(s) would you use for the following scenarios:

- changes in pebble sizes from one end of a beach to another
- examining the differences in the distribution of different ethnic groups within wards in a city
- how the concentration of PM_{10} particles changes with distance from the centre of an urban area
- examining the varying orientation of the long axes of drumlins in two areas of study
- examining the potential variation in sediment sizes of fluvioglacial deposits at two locations

Other aspects of number

It is clear from the preceding sections that an understanding of number is important in the application of any quantitative technique. The advantages of using numbers in geographical study are:

- they are precise and accurate (assuming the source is reliable)
- they can be collected easily through sampling methods and other forms of data collection methodologies

- they can be analysed statistically
- their collection and analysis can be replicated easily

There are, however, some disadvantages:

- their use is dependent on reliable sources and data collection methods
- they can reduce complex issues and opinions to simple numerical values
- complex analysis by statistical methods can produce simple numeric conclusions

It should go without saying that any measurements you undertake should be accurate, and that you are aware of the possibility of error. You should also appreciate the meaning of number, and how absolute values compare against each other. For example, when comparing demographic information for the developed and less developed world (Table 13), note how relative differences in birth rate and death rate can combine to create even larger differences in relative rates of natural increase.

Table 13 Extracts from the Population Clock, 2016

Indicator	World	Developed countries	Less developed countries	Approximate multiple difference (Less developed versus developed)
Births per minute	280	26	254	× 10
Deaths per minute	109	24	85	× 3.5
Natural increase per minute	171	2	169	× 85

Source: Population Reference Bureau

The understanding of number is also subject to change. Study the information in Table 14. It shows how levels of absolute poverty have decreased, and are likely to continue to decrease, in Asia between 1981 and 2030.

Table 14 Extreme poverty (below US$1.25 a day) in Asia (1981–2030)

Date	Poverty rate (%)	Numbers living in poverty (billions)
1981	69.8	1.59
2005	26.9	0.90
2015	12.7	0.47
2025	2.5	0.10
2030	1.4	0.06

Source: World Bank

It would appear that absolute poverty in Asia has fallen and will continue to fall, significantly, both proportionately and numerically. By 2030, the poverty of living on less than US$1.25 a day could largely be eradicated in Asia. But, does this statement hide the problem some countries still face? Views differ on how much money is needed to escape absolute poverty. In 2011, the World Bank proposed raising the poverty line from US$1.25 a day to US$1.78 a day. This is a change in the baseline. This higher value is likely to increase the number of people in extreme poverty in Asia by almost one-third.

Finally, another aspect of understanding number is to recognise the value of indices — where changes in number can be compared with a base figure. Figure 27 shows how house prices have changed over time since 1980 in five developed nations. Not only can you see how prices have increased over time, but also the relative rates of change in the five countries can be understood from the different gradients in the lines.

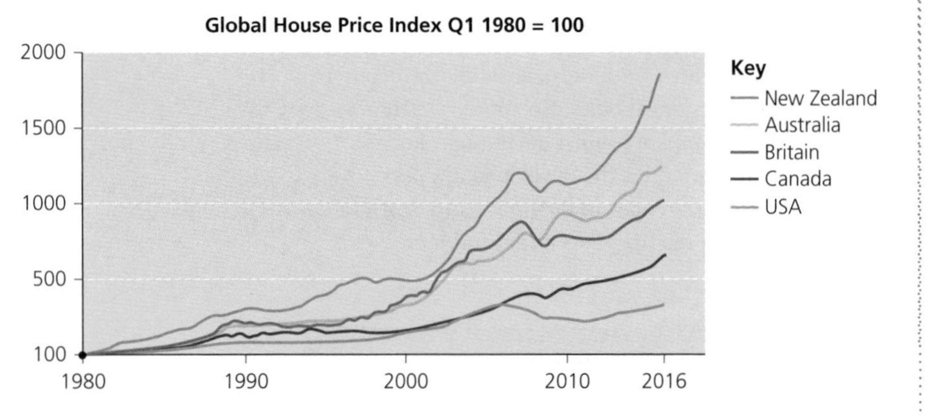

Figure 27 Changes in the global house price index (selected countries) 1980–2017

Summary

After studying this section, you should:

- be aware of the specification requirements of the geographical skills that need to be understood at both AS and A-level
- understand how to use and apply a range of qualitative techniques such as interviews, visual sources of data and coding, and be aware of the ethical dimensions of the enquiry process
- understand how to use and apply a range of cartographical and graphical techniques, including those associated with geospatial and geo-located technologies, in a range of contexts
- understand how to use and apply a range of statistical techniques such as measures of central tendency, dispersion, the Gini coefficient, the Student's *t*-test, Spearman's rank correlation and the Chi-squared test in a range of contexts
- be aware of the importance of number when studying data

Fieldwork

Fieldwork at AS

Specification requirements

Students of AS geography must undertake a minimum of **two** days of fieldwork and your centre will be required to provide evidence in the form of a written fieldwork statement that you have had appropriate opportunities for meaningful research. Students should follow the route to enquiry (see Figure 28) and be fully engaged in the decision-making processes in relation to the fieldwork and research.

AS fieldwork is externally assessed in AS Paper 1 and AS Paper 2. The specification requires AS students to carry out fieldwork in relation to physical and human geography:

Area of study 1 Topic 2: Glaciated landscapes and change **OR** Coastal landscapes and change

AND

Area of study 2 Topic 4: Regenerating places **OR** Diverse places

Through your fieldwork and research, you will be able to develop and demonstrate the full range, variety and diversity of fieldwork skills required. The skills you are required to demonstrate at AS are:

1. identify appropriate field research questions, based on your knowledge and understanding of relevant aspects of physical and human geography
2. undertake informed and critical questioning of data sources, analytical methodologies, data reporting and presentation, including the ability to identify sources of error in data and to identify the misuse of data
3. understand how to observe and record phenomena in the field and be able to devise and justify practical approaches taken in the field (including frequency/timing of observation, sampling and data collection approaches)
4. demonstrate knowledge and understanding of how to select practical field methodologies (primary) appropriate to your investigation
5. demonstrate knowledge and understanding of implementing chosen methodologies to collect data/information of good quality that is relevant to your topic of investigation
6. demonstrate knowledge and understanding of the techniques appropriate for analysing field data and information and for representing results, including GIS, and show ability to select suitable quantitative or qualitative approaches and to apply them
7. apply existing knowledge and concepts to identify, order and understand field observations
8. show the ability to present and write a coherent analysis of fieldwork findings and results in order to justify conclusions, as well as to interpret meaning from the investigation, including the significance of any measurement or other errors

General advice

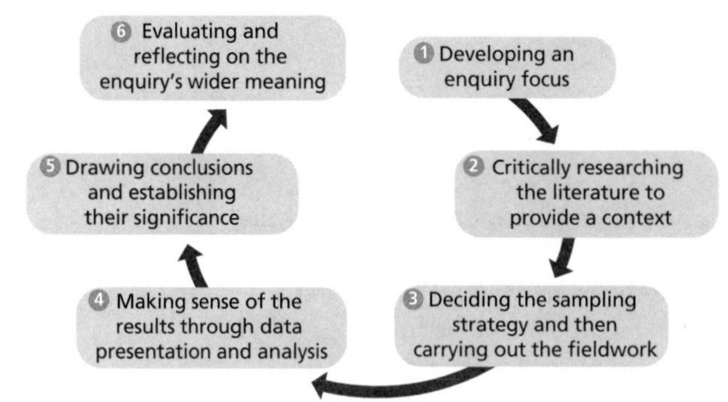

Figure 28 The route to enquiry

The geographical route to **enquiry** (Figure 28) is at the centre of good **fieldwork** and research. It is essentially a set of processes that allows an entire enquiry to be completed from start to finish. It consists of a series of linked individual stages that build together and allow reflection from one stage to the next. This set of stages should culminate in a complete enquiry, though there is no requirement to formally write it up.

Exam tip

There is no need to individually write up a full piece of fieldwork for the examination as you do not need to hand it in and you won't be assessed on it in its entirety.

Enquiry is the process of investigation (through a sequence of stages) to find a probable (or plausible) answer to a question(s) or aim(s) that has been developed.

Fieldwork is any work carried out in the outdoors. All AS geography students must undertake two days of fieldwork that include physical and human geography.

The examination

Although you need to undertake a full enquiry process as part of the AS course, the AS examination is focused on four optional themes linked to:

1 Glacial landscapes and change (physical)
2 Coastal landscapes and change (physical)
3 Regenerating places (human)
4 Diverse places (human)

You will be required to answer one question from the physical pair (1 or 2) and one question from the human pair (3 or 4). **It is therefore expected that you undertake one piece of fieldwork in either a glacial or a coastal area, and one piece of fieldwork in a place that will demonstrate either regeneration or diversity.**

The **fieldwork questions** are derived from the topics within the specification content of these options. The questions can be taken from any of the stages of the enquiry process, but probably not across all of them in any one paper. The fieldwork questions will assess:

- your knowledge and understanding of investigating geographical questions and issues
- your interpretation, analysis and evaluation of fieldwork data relating to an unseen fieldwork context
- your ability to construct arguments and draw conclusions in relation to your own fieldwork experience.

A **fieldwork question** is something that you can be asked in the context of the overall aim/focus of the enquiry.

Exam tip

Keep a field notebook of your fieldwork. It can be used as a key revision tool before the exam, helping you to remember the sites you visited and exactly what you did, and how and why you performed particular activities.

Exam tip

You might want to work in a group to create a concise set of notes to be shared before the examination.

There are 18 marks for each of the specific fieldwork sections. Questions will be set on materials that are both **unfamiliar** (9 marks), and **familiar** (9 marks) to you. The former will be based on new data provided to you on the day of the examination. You will be expected to process these data, which might include presentation, analysis and evaluation. You will be expected to use and interpret any of the skills we have looked at in the first part of this book in relation to these 'unfamiliar' data. The latter section will be on general field techniques (such as sampling methods and methods of data collection, presentation and analysis) and the fieldwork you have undertaken.

Unfamiliar data are data provided unseen in the examination context.

Familiar data are data that you (or your group) collected on your fieldwork.

Exam tip

You can record your field notes orally, for example using a recording app on your smartphone.

Assessment objectives

For AS and A-level geography, all assessments will test one or more of the following Assessment Objectives (AOs):

AO1: Demonstrate knowledge and understanding of places, environments, concepts, processes, interactions and change, at a variety of scales.

AO2: Apply knowledge and understanding in different contexts to interpret, analyse and evaluate geographical information and issues.

AO3: Use a variety of relevant quantitative, qualitative and fieldwork skills to: investigate geographical questions and issues; interpret, analyse and evaluate data and evidence; construct arguments and draw conclusions.

The 18-mark fieldwork-related questions all assess AO3.

Exam tip

Remember that any of the geographical skills (e.g. graphical, cartographical, statistical) can also be assessed elsewhere in the examination papers.

The enquiry focus

To prepare for your fieldwork activities, you should begin to pose geographical questions and/or devise suitable **hypotheses**, consider appropriate primary data collection methodologies and design survey strategies. You should think about selecting appropriate raw data/information (primary data) to be collected in the field, both qualitative and quantitative. This may involve taking measurements and carrying out surveys (for example, questionnaires, observations and interviews) and making images, including field sketches and photographs. You should also consider appropriate sampling techniques. In order to understand the theoretical or comparative context of your research question(s), you should read and/or collect secondary data/information. After your various fieldwork activities, you should then consider appropriate methods of data/information presentation and interpretation, analyse patterns and trends, draw conclusions and evaluate techniques and results.

A **hypothesis** is a statement or idea that can be tested by carrying out fieldwork.

Exam tip

Prepare a cartographic, graphical and statistics revision booklet that justifies, as well as describes, the various techniques you used on your quantitative data. You will likely need these techniques for the questions based on 'unfamiliar' fieldwork data.

Here are some examples of the type of questions based on your 'familiar' fieldwork.

- Explain the reasons why the location chosen was suitable for the investigation.
- Assess the accuracy of your methods to collect primary data.
- Describe how you collated your data and then presented them.
- Evaluate the reliability of the secondary sources you accessed.
- Justify why you chose one particular method of analysing your data.
- How valid were the conclusions that you drew from your results?

Exam tip

Prepare another list of the types of questions you could be asked regarding the familiar aspects of your fieldwork. Practise responses using sample exam questions and work with mark schemes.

You will have noticed that some other key words are used in both the description of the enquiry focus above, and in the sample examination questions. Several of these are explained below.

Primary data: fieldwork data that you collect yourself (or as part of a group) are called primary data — first-hand information that comes from you and people you have worked with.

Secondary data: information that someone else has collected, for example another person, group or even another organisation. Secondary data are very important in providing background information and a context for the enquiry — they may form part of your literature research for example. Secondary data can be:

- numerical: weather data, census data, deprivation data, crime data, river discharge data
- written: blogs, diaries, research reports, newspaper or magazine materials
- visual: maps, plans, photographs, satellite images, video

There is an important difference between primary and secondary data but, as technology progresses, there is a blurring between the two types. Consider data that might be collected as raw and unprocessed census information, personal opinion on YouTube, or crowd-sourced 'big data' from Twitter. These examples don't fit easily in either category and are sometimes called **tertiary** or **hybrid data**.

Conclusions should:

- summarise the main geographical outcomes that have been discovered, including any evidence (**qualitative** and **quantitative**) that backs them up
- relate findings back to the aim/focus of the investigation
- establish geographical links between factors and processes that have been uncovered as part of the fieldwork, indicating where there are weak and strong linkages
- accept that geography can produce 'messy' outcomes in which the results reveal a complex, unfamiliar and sometimes unpredictable pattern
- begin to comment on the geographical significance of your study and whether the results challenge similar studies or are, in fact, in broad agreement with them

Conclusions must always be brief, factual and to the point. This is important when linking them to a question in the examination.

Qualitative data are non-numeric data, such as people's views and opinions, photographs and images.

Quantitative data are precise, numerical data which can be analysed statistically. Furthermore their collection can be replicated.

The final part of the route to enquiry is reflection. Critical reflection is much more than a list of what did (or did not) go well. It is an opportunity to review the whole enquiry process, which may include reference to the fieldwork results as well as the secondary research information. Importantly, this stage represents an opportunity to comment on the **accuracy**, **reliability** and **validity** of the results; in other words, how much they can be trusted under different circumstances.

You should include a brief analysis of how the process could be improved, for example in terms of repeating particular procedures or using different techniques, including those linked to data presentation and analysis. Try to avoid simplistic and obvious statements in the fieldwork exam (which can sometimes read like excuses). Instead, think carefully about any possible sources of error that may have been introduced through the equipment, sampling method or operator limitations.

Accuracy is the degree of closeness of a measurement (or method) to the actual value of the item (or process) being measured.

Reliability is the consistency of a measurement or source. A source is said to be reliable if it produces materials that can be trusted over a long period of time.

Validity is the extent to which the outcomes actually reflect the true characteristics of the feature(s) being measured, i.e. are they well-founded?

Using geospatial data for place-related studies

For fieldwork relating to Regenerating places or Diverse places, you could consider using online sources for socio-economic data. Some sources include:

- The Office for National Statistics (ONS): https://www.ons.gov.uk
- DataShine maps (based on the 2011 census): http://datashine.org.uk/
- Consumer Data Research Centre (CDRC) maps: https://data.cdrc.ac.uk
- Index of Multiple Deprivation (IMD) (see below): http://apps.opendatacommunities.org/showcase/deprivation/

It is recommended that you spend some time investigating what can be found on these sites. Each shows choropleth maps with additional overlays of spatial information, such as settlements, rivers, roads, railways. Note that the more detailed DataShine and CDRC maps present data by super output area (SOA), whereas the ONS mapping tool uses data aggregated by ward.

The Index of Multiple Deprivation (IMD) 2015

The Index of Multiple Deprivation (IMD) for England is published by the Department for Communities and Local Government and informs national and local government decision-making and associated patterns of investment. It ranks the SOAs across the whole country according to a combination of seven domains of deprivation: income, employment, education, health, crime, barriers to housing and services, and living environment.

Each of these domains is based on a further number of indicators — there are a total of 37. As far as is possible, each indicator is based on the most recent data available although, in practice, most indicators in the 2015 data, for example, relate to the tax year ending April 2013.

The IMD uses smaller parts of the SOAs, called Lower SOAs (LSOAs). Each contains about 1,500 residents or about 650 households. Such a fine spatial framework is capable of identifying small pockets of deprivation.

The **deciles** shown on the map are produced by ranking 32,844 LSOAs and dividing them into ten equal-sized groups. Decile 1 represents the most deprived 10% of areas nationally and decile 10, the least deprived 10%. When interpreting these data, note:

- the rank of the deciles is relative: they simply show that one area is more deprived than another but not by how much

- there are large areas of colour, often the same or similar shade. Note that these show areas and not numbers of people living there
- the data shown by such neighbourhood-level maps provide a description of areas as a whole and not of individuals within those areas
- deprivation, not affluence, is mapped

A number of sample AS examination fieldwork questions, and answers to them, can be found on pp. 93–107. They cover a range of skills and fieldwork contexts.

Fieldwork at A-level

Specification requirements

A-level students should complete a minimum of **four** days of fieldwork. This fieldwork must relate to processes in **both** physical and human geography. It must also provide an introduction to the nature and process of a high-quality geographical enquiry. Your centre will provide evidence of this fieldwork (which can include days undertaken as part of a separate AS in geography) in a written fieldwork statement.

The fieldwork will enable you to develop skills that you can use in your independent investigation. You can, if you wish, use data you collect in your fieldwork as part of your independent investigation. However, you can also carry out an independent investigation on a separate topic and collect new data.

Coursework: independent investigation

The purpose of this coursework is to test students' skills in independent investigation that involves (but which need not be restricted to) fieldwork. Your independent investigation may relate to human or physical geography, or it may integrate them. Your independent investigation must:

- be based on a question or issue defined and developed by you individually to address aims, questions and/or hypotheses relating to any of the compulsory or optional content in the specification
- incorporate field data and/or evidence from field investigations, collected individually or in groups
- draw on your own research, including your own field data and, if relevant, secondary data sourced by you
- demonstrate that you have independently contextualised, analysed, evaluated and summarised your findings and data in your report, which should be 3,000–4,000 words
- show that you have drawn your own conclusions and communicated them by means of extended writing and the presentation of relevant data

You can complete your independent investigation report at school/college, at home (or other location outside school/college), or at a combination of both.

The total number of marks for the independent investigation is 70 and it is worth 20% of the A-level qualification.

The specification provides guidance on where independence is required, and where collaborative work (with other students) is allowed. This is summarised in Table 15.

Table 15 Investigation stage and required level of independence

Investigation stage	Level of independence
Exploring the focus	Collaboration allowed
Title of the investigation, focus of investigation, purpose of the investigation	Independent work
Devising methodology and sampling framework	Collaboration allowed
Primary data collection	Collaboration allowed
Secondary data collection (if relevant)	Independent work
Data/information presentation	Independent work
Data analysis and explanation/interpretation	Independent work
Conclusions and evaluation	Independent work

The specification also provides a table (Table 16) which shows a suitable A-level geography route to enquiry for the independent investigation.

Table 16 A route to enquiry

Stage	Description
Purpose, identification of a suitable question/aim/hypothesis and developing a focus	Identify appropriate field research questions/aims/hypotheses, based on knowledge and understanding of relevant aspects of physical and/or human geography. Research the relevant literature sources linked to possible fieldwork opportunities presented by the environment, considering their practicality and relationship to compulsory and optional content. Understand the nature of the current literature research relevant to the focus. This should be clearly and appropriately referenced within the written report
Designing the fieldwork methodologies, research and selection of appropriate equipment	Consideration of how to observe and record phenomena in the field and to design appropriate data-collection strategies taking account of sampling and the frequency and timing of observation. Demonstrate knowledge and understanding of how to select practical field methodologies (primary) appropriate to the investigation (which may include a combination of qualitative and quantitative techniques)
Information collation and data representation and analysis	Know how to use appropriate diagrams, graphs, maps and geospatial technologies to select and present relevant aspects of the investigation outcomes
Analysis and explanation of information	Use techniques appropriate for analysing field data and research information. Demonstrate the ability to write a coherent analysis of fieldwork findings and results linked to a specific geographical focus
Conclusions and critical reflection on methods and results	Use knowledge and understanding to interrogate and interpret meaning from the investigation (theory, concepts, comparisons) through the significance of conclusions. Demonstrate the ability to interrogate and critically examine field data (including any measurement errors) in order to comment on its accuracy and/or the extent to which it is representative and reliable
Recognising the wider geographical context	Explain how the results relate to the wider geographical context and use the experience to extend geographical understanding. Show an understanding of the ethical dimensions of field research

The rest of this section examines the requirements of the A-level independent investigation in more detail, and follows the sequence of the official mark scheme as provided in the specification.

Purpose of the independent investigation

This is worth **12 marks** and the assessment criteria state that the following will be assessed:

(a) Geographical knowledge and understanding of location, geographical theory and comparative context (AO1)

(b) Understanding of links between the investigation's context and a broader geographical context (AO2)

(c) The use of sources in order to identify/obtain geographical information and data that support the investigation. How the research information is used to construct an aim, question or hypothesis in providing a framework for investigation that is manageable. The structure of the enquiry process (AO3)

When assessing the NEA, the assessor will make a judgement of performance for each criterion and then apply a 'best fit' judgement for all three and allocate a level for the whole section. Marks will be awarded from within the range of marks for that level. There is no weighting of the criteria — they are equal.

What you should do for each of these outcomes

(a) Geographical knowledge and understanding of location, geographical theory and comparative context (AO1)

(b) Understanding of links between the investigation's context and a broader geographical context (AO2)

Fieldwork in geography has a long tradition. It is seen as an opportunity to discover and explore, as well as to test geographical ideas. At the heart of fieldwork lies the route to enquiry (see Figure 28 on p. 50). This constitutes a series of stages in developing a geographical enquiry from start to finish. It can be summarised in four simple steps:

1 creating a need to know
2 using data that have been collected
3 making sense of those data
4 reflecting on the outcomes and the activity

More detail is provided in Table 16 above.

Your investigation must have a geographical meaning. It should be rooted in one or more places. It must have a purpose and you should have some sense of where it is heading, or what you expect/hope the outcomes to be. Having developed an idea, you could conduct research into relevant literature. This will provide additional background and a theoretical and locational context. You may be able to explore similar places, obtain up-to-date thinking on the topic and read opinions that have previously been voiced about your chosen issue.

Therefore, your independent investigation should:

- involve **research of relevant literature sources**
- demonstrate **understanding of the theoretical** and/or **comparative context** for your research question, hypothesis or issue
- demonstrate good practice in terms of **referencing** and using a **bibliography system**

Teachers should ensure that you have access to relevant literature sources, including textbooks, magazines and journals, other written material stored within the geography department, and any other external sources such as library and online facilities. It is important to start collecting resources early about both the specific area and the general theme chosen. Any work you do here should be clearly and appropriately referenced in the written report. A consistent referencing style should be used (see box).

Systems of referencing

- **Book:** name(s) of the author(s); the year published; title; publisher; pages used.
- **Newspaper:** name(s) of the author(s); the year published; article title; newspaper name.
- **Online source:** name(s) of the author(s); the year published; page title; website name; 'available at:' URL; the last date you accessed it.

It is important that you conduct your investigation at an appropriate geographical scale. This will of course depend on the precise nature of your investigation. However, it is likely that to obtain the necessary primary data, there will need to be some work undertaken at a local scale.

(c) The use of sources in order to identify/obtain geographical information and data that support the investigation. How the research information is used to construct an aim, question or hypothesis in providing a framework for investigation that is manageable. The structure of the enquiry process (AO3)

Your independent investigation must:

- be based on a research question, hypothesis or issue **defined and developed by you individually**
- relate to **any part of the specification** content

You may discuss general ideas and research for appropriate geographical questions with your fellow students and with your teacher/field study tutor. Following this initial stage you must then finalise the focus of your investigation and draft a title by yourself — **independence is crucial**. See below for advice.

In the Edexcel geography independent investigation form and final written report, you must provide a clear justification and contextualisation of how your enquiry will help you to address your title and explore your theme in relation to your chosen geographical area. Your teacher should review your independent investigation proposal to make sure your proposed investigation fulfils the specification requirements and that there is sufficient scope for you to access the full range of

Exam tip

You are not allowed to choose from a list of titles or investigations provided by your teacher or a field study centre. However, other sources are available.

marks available for the NEA. Your teacher will approve your investigation proposal once they can be sure that you have independently devised your own hypotheses and/or questions and/or sub-questions, even though the title may be the same as/similar to another student's.

Your teacher/field study tutor should advise you on **health and safety** considerations. These are now well-established in schools and colleges, but you may wish to adhere to the advice, shown in the box, from the Field Studies Council (FSC).

FSC health and safety advice

Risk assessment is the fundamental tool to ensure safety is effectively managed. The purpose of the risk assessment process is to identify hazards; assess who may be harmed and how, and manage the hazards through safe systems of work. In line with Health and Safety Executive (HSE) guidelines, you should follow five steps to risk assessment:

1 identify the hazards
2 decide who might be harmed and how
3 evaluate the risks and decide on precautions
4 record your findings and implement them
5 review your assessment and update if necessary

The likelihood and severity of the hazard(s) occurring can be scored numerically (1 = low, 5 = high), with resultant risk being assessed as:

- >10: take immediate action to either remove or control the risk — for example take a less risky option, or prevent access to the hazard
- 8–10: inform people of the risk and look at ways of reducing it
- <8: monitor the situation closely and aim to reduce risk over the longer term

All significant findings should be documented and periodically updated unless changed circumstances dictate an earlier review.

Health and safety is important. Prior to any work 'in the field', safety needs to be considered and a full risk assessment may be needed before any field trips can take place. An example of a risk assessment table is given in Table 17. Any risk assessment first requires identification of actual or potential hazards and then an indication of how these can be overcome or reduced to an acceptable level. Your teacher may want to discuss the safety aspects of your investigation with you.

Some safety issues are very obvious, especially when working in exposed and remote physical environments, but the less obvious or less extreme risks also need to be considered. For example, when working in a coastal location the actual risk of drowning is likely to be very small indeed, whereas slipping on rocks and twisting an ankle is much more likely. In an urban environment the dangers of traffic may be obvious but the need to think about how to carry out interviews and avoid the risk of being isolated, or receiving verbal or other abuse, should be important in your planning. **Make sure what you do is safe, and that others know where you will be.** Be fully aware of the steps you have taken to keep yourself safe.

Table 17 An example of a table of risk

Risk/hazard	Who might be involved	Level of risk	Precautions
Getting caught by waves when measuring beach characteristics	Myself and other students	1 = low	Always collect data from at least 5 m above the swash zone

More advice is given in Table 18.

Table 18 Examples of actions that should be considered to keep you safe during fieldwork

Let someone know where you are	Always tell someone where you are going and when you are expected back, especially when working alone
Check mobile phone reception	Don't assume you will always get mobile phone reception — check that first. One useful tip in an emergency: if you are using a car, raise the boot and stand behind it. It can act as a reception disc for a weak signal
Have the right equipment	If working in a remote area, carry a first aid kit and a torch. If working below a cliff, wear a hard hat
Carry a map — paper or electronic	Stick to identified paths in rural areas. Follow way-marked paths if they exist
Traffic	Always walk on the side of the road facing on-coming traffic. Wear a high-vis jacket
Wear appropriate clothing and footwear	This is particularly relevant when working in cold and/or wet weather

Exam tip

Risk awareness and risk assessment do not form part of the assessment process, but it is essential that they are undertaken.

Aim, research question, hypothesis or issue?

The **aim** of an investigation is what you are generally trying to achieve in your fieldwork location. This will depend on time, location, environmental conditions, equipment available and risk assessments. For example, your aim might be:

To study the changes in infiltration rates over time in drainage basin X

or

To study the effectiveness of traffic-calming measures in town X

Alternatively, you might want to express what you are investigating as a **research question**:

What factors influence infiltration rates in drainage basin X?

or

To what extent are golf courses an environmental, economic or social asset in rural area C?

or

How has gentrification changed the character of place D?

For the last of these, you could also think about breaking down the overall question into smaller sub-questions. You could have two or three such sub-questions, such as:

Exam tip

If you decide to have a series of sub-questions they must be closely tied together into a single theme or focus, otherwise the investigation could become too large.

What social and demographic changes have occurred in place D in recent years?

What have been the impacts of these changes on housing and services within place D?

What are the attitudes of people to the changes that have occurred in place D?

You may wish to test one or more hypotheses. A **hypothesis** is a statement based on a question which can be either proved or disproved, often involving statistical testing, such as:

A number of factors cause flooding to occur at P

A range of management strategies, natural and man-made, are used to protect area P from flooding

Everyone thinks the flood management strategies at P are effective

Shingle beaches have a steeper gradient than sand beaches

You may wish to **evaluate an issue** in a local area. For example:

The plan to build a Tesco Local store in place J has created a range of attitudes amongst the local people

The increase in atmospheric pollution in town K is unpopular and potentially unhealthy

The proposed development of new housing in the suburban area of L is unnecessary

In choosing the **location** for your investigation, you should consider: accessibility; safety; availability of appropriate equipment and resources, including data availability and manageability.

It is essential to establish that a piece of research work is feasible at the outset. Therefore it may be worth carrying out a simple **feasibility study** or **pilot survey** to make sure that you do not waste your time and that the information you require is available to you. Some hypotheses may be impossible to test because of practical problems of measurement (e.g. rates of erosion or mass movement on river banks or coastal cliffs), inaccessibility (e.g. research locations on private property) or lack of secondary sources that can be freely accessed. For example, some historic river and catchment data from the National Rivers Flow Archive (NRFA) cannot be easily accessed without special permission. If conducting a questionnaire you could try out the questions on a small trial group to see if you get the outcomes you wanted. After this you can then modify the technique by tweaking the questions.

Problems can also arise as a result of boundary changes to the small spatial units used to aggregate census population data. In 2001, for example, the old enumeration districts for small areas were replaced by a series of new 'super' and 'output areas' (see p. 53). This lack of comparable spatial data would make it extremely difficult to attempt an investigation of population change in a suburb or market town for, say, the period 1971 to 2011.

You should document and present evidence of a pilot survey in your final report. It demonstrates that you are reflective, that your design is flexible and adaptable, and that you have thought and planned.

Some, or all, of your work must be undertaken outside the classroom in the field, such that **first-hand (primary) data** have to be collected. This may, of course, include work undertaken at a field study centre or in a work-experience setting. This work can be undertaken independently or as part of a group. Any data you collect from other written sources are classed as **secondary data**, and they must be collected independently. If your investigation involves some form of comparison with another location, or another area of study, then the **comparative context** (the similarities and differences between the areas of study) must be made clear in your write-up.

Field methodologies and data collection

This is worth **10 marks** and the assessment criteria state that the following will be assessed:

(a) Methods to collect data and information relevant to the geographical topic (AO3)

(b) A sampling framework linked to the geographical topic being investigated (AO3)

(c) Consideration of frequency and timing of observations (AO3)

(d) Research planning that shows understanding of the ethical dimensions of field research methods (AO3)

(e) Reliability/accuracy/precision of data and information collection (AO3)

When assessing the NEA, the assessor will make a judgement of performance for each criterion and then apply a 'best fit' judgement for all five and allocate a level for the whole section. Marks will be awarded from within the range of marks for that level. There is no weighting of the criteria — they are equal.

What you should do for each of these outcomes

All of these outcomes concern the methodologies required to **collect** data for the investigation. However, there are other considerations that are implied when carrying out these methodologies:

- the difference between primary and secondary data
- identification and selection of appropriate physical and human data
- measuring and recording data using different sampling methods
- description and justification of the various data collection methods

Your investigation must incorporate the observation and recording of field data (primary data) that should be of good quality and relevant to the topic under investigation. It should involve justification of the practical approaches adopted in the field including frequency/timing of observation, sampling strategies used and data collection approaches. Your outcomes should draw on your own research, including your own field data and/or secondary data, and your experience of field methodologies of the investigation of core human and physical processes.

You may collaborate when planning and selecting your methodologies or sampling strategies. One important element of the teacher approval process of your investigation proposal is to ensure that you have made use of appropriate methodology and sampling strategies. If you do not fully justify your methodology and sampling, you may limit your access to marks.

Primary data collection may be carried out individually or in groups, but there must be evidence of your own collection of data in the investigation. The specification strongly suggests that students make use of both **quantitative** and **qualitative** data.

Secondary data collection (if relevant) must be carried out independently — you should select your secondary sources of data on your own. The use of both quantitative and qualitative data should also be considered here.

Primary data are information collected by you for the first time through a personal field investigation.

Secondary data are information derived from published documentary sources that has been processed, such as processed census data, research papers, text books, internet, etc.

Quantitative data are data in numerical form which can often be placed into categories and analysed statistically.

Qualitative data are non-numerical data such as photographs, sketches and may involve the collection of opinions, perspectives and feelings from questionnaires and interviews.

Sampling

Prior to the investigation, you should consider the various forms of sampling, the most common ones being:

- random sampling
- systematic sampling
- stratified sampling

See p. 12 for more detail on these forms of sampling. Collecting a 'good' sample is integral to the design of the investigation — the more rigorous and reliable the methods of data collection, the greater the validity of the conclusions.

There are three key aspects of sampling:

1 deciding on an appropriate sample area (or frame)

2 choosing an appropriate sample size

3 selecting the best sampling method

The **sample area** (**frame**) is linked to what you know and understand about the population under investigation, for example, a grid placed over a particular residential area or a transect lines across a rural landscape. The precise characteristics of the sampling frame should be established, along with any unusual patterns, features or characteristics that might be avoided. This will reduce sampling bias. The area should also be defined as a discrete geographical area, for example a section of coast, part of a catchment or two census output areas within two different parts of an urban area.

As most natural populations in geography are very big (and spread over a large area or change over time), investigations are inevitably based on sample data. Decisions have to be made on the quantity of data to collect (i.e. the **sample size**) and objective sampling methods to ensure representativeness. The choice of scale is also a significant decision. An enquiry into the spatial pattern of consumer spending in a central shopping area requires a town or city large enough to have at least five or six clearly defined shopping streets/areas. A large town or city of 300,000 people is likely to be much more suitable than one of 30,000.

In terms of sample size, think about:

- what is the *minimum* number of data items that you need to collect so that you can carry out analysis (including statistics) reliably?
- what is the *maximum* number of data items above which the results do not change? In other words, at what point will collecting more data not significantly improve the accuracy of results?

Deciding on the correct **sampling method** requires some knowledge of the population. In reality, constraints of time and resources impose limits on geographical investigations (although these should not be used as 'excuses'). Most investigations must strike a balance between the amount of data collected and the time and resources expended on data collection. Considerations connected to time and resources are essentially practical. Several key questions must be considered alongside the fieldwork design and methodology.

- How much time will be devoted to data collection and how much potential fieldwork time will be spent travelling (and walking) to and from the fieldwork site?

Exam tip

You should not use primary or secondary data that were collected by someone other than yourself or you as part of a group. This includes parents and teachers.

- Will data be collected by a group or by an individual?
- If a group, how many students will be needed for the data collection process? This last consideration is important in questionnaire surveys, where rejection rates are often high and where each interview may take several minutes.

Equipment

An important aspect of data collection is equipment, especially in physical geography where fieldwork often involves data collection through measurement. Slope surveys, for instance, using instruments such as clinometers, ranging poles and tapes, can generate numerical data on valley and beach profiles. Equipment including anemometers, thermometers and humidity meters are used to measure microclimate. Soil pH is measured using a BDH soil-testing kit; and callipers measure the long, short and median axes of sediment.

In all cases of fieldwork, a well-laid out and logical recording sheet makes data processing much more straightforward at a later date.

Questionnaires and interviews

These were discussed on p. 7. There are several ways of collecting these qualitative and semi-quantitative data: on-the-street surveys, which are most often conducted in public spaces (e.g. a shopping street), though occasionally doorstep interviews are undertaken; and remote questioning (e.g. a postal or online survey). In each case, the interview or questionnaire will need to be customised for both the target audience and the mode of delivery. Questionnaires can sometimes fail because of poor design, questioning sequence or if the questions are poorly constructed.

Ethical dimensions of field research

These have also been discussed earlier in this book — see p. 14.

Collecting appropriate data

The assessment statements given above use various descriptive words: **appropriate**, **reliability**, **accuracy** and **precision**. Collecting inappropriate and/or unnecessary data does not assist you. Your initial planning will have determined a key question or hypothesis based on part of the specification content. Discussions with your teacher and/or fellow students should then focus on what data are required to address the key question/hypothesis. This will result in a range of possible data being identified, some of which may have no relevance to the task, and then selection of the most **appropriate** data can take place and methods for collecting these data can then be determined.

It is important to collect data that are both precise and accurate, and hence by definition, reliable. The **precision** of a measurement is the degree to which repeated measurements under unchanged conditions show the same results. The **accuracy** of a measurement is how close each measurement comes to the real value. The further a measurement is from its expected value, the less accurate it is. **Reliability** refers to the consistency or reproducibility of a measurement. A measurement is said to have a high reliability if it produces consistent results under consistent conditions. True reliability cannot be calculated — it can only be estimated based on knowledge and understanding of the topic.

Finally, you will be required to describe and justify your data collection methods. You must understand **how** you collected your data. This is likely to be presented as a written step-by-step approach. The probable success of your fieldwork will be greatly enhanced by being very clear as to how your data were collected. However, a higher level outcome is your **justification** of the methods. You should understand **why** each aspect of your data collection methods was carried out. By being prepared to be asked, and answer, evaluative questions on your data collection methodology, you should be able to justify it clearly.

Exam tip

Justification concerns not just why a particular methodology has been selected but also why other methodologies have been rejected.

Further points

As suggested earlier in this book, there is a strong push for you to use data from online sources, sometimes called **'big data'**. With the increase in availability of such sources of data, many of which have been unprocessed, there is some debate amongst geographers that the traditional distinction between primary and secondary data is out of date. Some argue that unprocessed data collected by you in this format are a form of primary data — they are only secondary once someone else has processed them. Certainly, crowd-sourced data from an outlet such as Twitter could be collected by you for the first time. Some have referred to this type of data as 'hybrid data' because:

- the methodology used to collect the opinions is similar to primary data collection
- the data themselves are 'raw' — they have not been collated and analysed before you collect them — again similar to primary data
- the opinions were pre-existing and were not generated as part of your investigation, and so are a secondary source

'Big data' are extremely large, dynamic, online data sets that may be analysed to reveal patterns, trends and associations, especially relating to human behaviour and interactions.

This is, however, a debate for professional geographers, and you should make sure that you collect some primary data in a fieldwork context. Note also that both quantitative and qualitative data can be collected within the online context.

Secondary data need to be used with care because:

- the data were often collected for a different purpose
- the methodology used may not always be clear
- sample size, and thus reliability, might not be clear
- the data may contain errors and these may not be obvious
- the data may be old and out-dated

Secondary data sources are very varied. They can be grouped into three main types — statistical, graphical and written (Table 19).

Table 19 Types of secondary data source

Statistical	Graphical	Written
Weather data	Maps and plans	Newspapers
River discharge data	Graphs and charts	Diaries
Census data	Satellite images	Radio, TV
Crime statistics	Photographs	Online sources
Deprivation data	Artistic works	Blogs and social media

Remember that an important part of using secondary research sources effectively is recording information about the source. During research it is easy to forget to do this (especially when using the internet), but it is important because:

- the source may need to be found and accessed again, to check details or get more information
- sources may need to be stated (cited) in the fieldwork report or the bibliography

As stated above, you should make sure you use both qualitative and quantitative data. Each of these types of data have their strengths and weaknesses (Table 20).

Table 20 Strengths and weaknesses of quantitative and qualitative data

Data	Strengths	Weaknesses
Quantitative	Precise, numerical Reliable as a result of sampling design Can be analysed statistically Collection can be replicated	Poor collection methods can lead to weak conclusions Reduces complex situations and views to numbers Complex analysis can produce simplistic mathematical outcomes
Qualitative	People's views and opinions provide a human perspective to numerical data Can suggest new research possibilities based on comments made	Can take a long time to collect Analysis can be difficult and outcomes may be tenuous Data are subjective and may not be reliable

You could include neat hand-drawn sketches and maps in your report, as well as labelled photographs to show how any complex equipment was used and how you maintained reliability and accuracy. If you are using a questionnaire, consider including an annotated blank form to show the reasons why you chose those questions and why you put them in the order in which they appear.

When you are describing your methodology(ies), consider using tables to summarise your work. Tables 21 and 22 are examples based on collecting primary data and secondary data, respectively. Remember you should also indicate which techniques were carried out on your own and which as part of a group.

Table 21 Primary data collection

Data source	Why used/ purpose	Method: when/ where	Justification of sampling type (if any)	Problems/ limitations	Improvements
Field measurements					
Land use survey					
Field sketch					
Photographs					
Questionnaire/ interview					

Table 22 Secondary data collection

Data source	Why used/ purpose	Method when/ where	Justification of sampling type (if any)	Problems/ limitations	Improvements
Government statistics					
Local area plan					
Local newspaper					
Websites					
Textbook					

Data representation, analysis, interpretation and evaluation of techniques and methodologies used

This is worth **24 marks** and the assessment criteria state that the following will be assessed:

(a) The use of geographical skills to deconstruct data in order to show connections and the statistical/geographical significance of data (AO3)

(b) Appraisal of techniques and methodologies including ethical dimensions of field research and utility and validity of the chosen methodologies (AO3)

(c) Synthesis of research findings to form conclusions (AO3)

(d) Communication of conclusions that are supported by the presentation of fieldwork data or information (AO3)

When assessing the NEA, the assessor will make a judgement of performance for each criterion, and then apply a 'best fit' judgement for all four and allocate a level for the whole section. Marks will be awarded from within the range of marks for that level. There is no weighting of the criteria — they are equal.

What you should do for each of these outcomes

All of these outcomes concern the methodologies required to **present**, **analyse** and **interpret** the field data collected during your investigation. However, much of the emphasis in these statements is on **analysis** of the field data and **interrogation** of that field data. Although you (and possibly your teachers) will consider that presentation is important in this process, remember that presentation is only a means to an end and it is the analysis that should take precedence. So, it is important that you apply, link or connect the outcome of the field observations as determined by the analysis to the underlying knowledge, theory or concepts that were introduced in the first part of the assessment process.

Exam tip

Analysis and **interpretation** are more important than **presentation**.

Before data and information can be represented in a meaningful way, it is likely that they will need to be collated, sorted and selected so that only the data and information that are relevant to the overall focus of the work are used. Tables can summarise complex information. They should be clear so that the information is easy for the person reading the work to understand. Use brief headings within tables and put the units in the column header. Giving row and column totals may make interpretation easier for anyone reading the work.

Data representation is about getting what you have found ready for analysis. At the initial stage of analysis, data are processed and presented as tables, charts and maps. Remember that the purpose of data representation is to clarify meaning, i.e. to make initial sense of what you have found.

You should select and use appropriate data presentation methods on your own. Similarly, you should select and use appropriate data analysis techniques and independently analyse and interpret the results on your own. You must understand **why** the method(s) selected are appropriate for the data concerned, and demonstrate the reasons for their selection and use.

There are other considerations that are implied when carrying out these tasks. There should be:

- an appreciation of a range of visual, graphical and cartographic methods
- selection and accurate use of appropriate presentation and analytical methods
- description, explanation and/or adaptation of appropriate presentation and analytical methods

When applying these presentation and analytical techniques, it is clear that you should use one (or more) qualitative and one (or more) quantitative technique. You could use some of the following in your investigation:

- annotation of illustrative and visual material: base maps; sketch maps; OS maps (at a variety of scales); diagrams; graphs; field sketches; photographs; geospatial, geo-located and digital imagery
- overlays, both physical and electronic
- factual text and discursive/creative material and coding techniques when analysing text
- numeracy — use of number, measure and measurement
- questionnaire and interview techniques
- atlas maps (only if relevant)
- weather maps — including synoptic charts (only if relevant)
- maps with located proportional symbols
- maps showing movement — flow lines, desire lines and trip lines
- maps showing spatial patterns — choropleth, isoline and dot maps
- line graphs — simple, comparative, compound and divergent
- bar graphs — simple, comparative, compound and divergent
- scatter graphs, and the use of best fit line
- pie charts and proportional divided circles
- radial charts
- triangular graphs
- graphs with logarithmic scales
- dispersion diagrams
- Lorenz curves
- measures of central tendency — mean, mode, median
- measures of dispersion — dispersion diagram, inter-quartile range and standard deviation
- inferential and relational statistical techniques, including Spearman's rank correlation, Student's *t*-test, Chi-squared test and the application of significance tests

Exam tip

Teachers/field study tutors must not suggest or provide guidance on specific methodologies for you to use.

- remotely sensed data
- electronic databases
- innovative sources of data, such as crowd sourcing and 'big data'
- ICT (GIS and online spreadsheets) to generate evidence of many of the skills provided above such as producing maps, graphs and statistical calculations

Data presentation and analysis go hand in hand, since the best way to present data is one that encourages or allows analysis and interpretation to take place. The first part of this book illustrates some of the many ways in which data can be accurately and meaningfully presented.

For presentation, bear in mind the following:

- include a wide range of appropriately chosen representation techniques
- in a geographical investigation, methods of presenting material spatially (i.e. with maps) will be important. These may be based on existing maps or be specially drawn by you for the purpose in mind
- simple techniques often work very well, such as using overlays or using a map as a base upon which to plot other information
- decide whether the data need spatial techniques or non-spatial techniques such as pie graphs
- photographs, preferably well labelled or annotated, are almost always useful
- computer graphics/geospatial mapping can help, and may be very attractive, but beware — you can overuse them and produce 'death by pie chart'

Exam tip

Do not use different techniques to show the same data. This serves no purpose and undermines your rationale for your choice of technique.

The presentation section is best integrated into an analysis or results section:

- line and scattergraphs are often very powerful aids, and when a trend line or a line of best fit is added they become analytical
- if you are involved in an investigation that is based on strict 'hypothesis testing' principles, don't forget the traditional and relatively straightforward techniques before you get stuck into deep statistical analysis. Remember a balance between quantitative and qualitative approaches is often best
- statistics should be used with a purpose. Difficulty in drawing a line of best fit on a scattergraph may suggest to you that you should apply a Spearman's rank correlation test to see if there is a valid correlation between your data sets

The first section of this book also provides details on a range of statistical skills including Spearman's rank correlation, the Student's *t*-test and the Chi-squared test. Other statistical tests exist that are not referred to on the Edexcel specification, for example Mann Whitney *U* or nearest neighbour analysis. You are at liberty to use these if you can, and if they are applicable — it may impress!

Exam tip

Quoting the final outcome of a statistical test is important, but you need to understand what it means in the context of your investigation.

Before using statistical tests:

- make sure that they are necessary, and that you have chosen the appropriate ones
- be sure you know how to interpret the results of the tests, taking into account the degrees of freedom for your data and significance levels, as appropriate

This part of the assessment is not just about doing and applying the presentation and analytical technique(s). You are also required to independently contextualise and summarise their findings and draw conclusions with regard to the knowledge, theory and concepts that underpin the investigation. This may begin with a descriptive summary of what your processed fieldwork data show. This should then be followed by both explanation and interpretation — where you offer the meanings emanating from your findings. These will eventually assist you in reaching a conclusion(s) of your investigation — see next section.

Finally, you should always be aware of **anomalies** in your fieldwork data which may become more apparent during the presentation and analytical processes. An anomaly might be viewed as being something that deviates from the norm, but for you it might, perhaps, best be described as being 'an irregularity' within any data you collect. It is often difficult to identify anomalies in raw data, unless the anomalies are particularly striking. However, once the data have been presented and processed, any anomaly or anomalies may become immediately apparent. Descriptions of, and possible explanations for, such irregularities should then be offered.

An **anomaly** is something which deviates from a trend, or from what is expected. On a scattergraph, anomalies are called residuals.

Anomalies could be explained by the complexity of geographical systems and features. Often their explanation involves recognition and understanding of more than one causal factor, as well as possible problems with your data-collection methodology and sampling accuracy.

Much of your analysis will be in terms of written description and explanation. You need to be clear and precise in your expression, quoting your evidence and justifying your identification of trends, correlations, relationships and anomalies. If done well, your presentation and analysis should run smoothly into your conclusions and overall evaluation.

Conclusions and critical evaluation of the overall investigation

This is worth **24 marks** and the assessment criteria state that the following will be assessed:

(a) Demonstration of geographical knowledge and understanding of location, geographical theory and comparative context (AO1)

(b) Understanding of links between the investigation's conclusions and a broader geographical context (AO2)

(c) Synthesis of research findings (AO3)

(d) Appraisal of the reliability of evidence and validity of conclusions (AO3)

(e) The use of argument, reasoning and the enquiry process. The use of geographical terminology (AO3)

(f) Conclusions involving the drawing together of evidence and concepts linked to the investigation (AO3)

When assessing the NEA, the assessor will make a judgement of performance for each criterion and then apply a 'best fit' judgement for all six and allocate a level for the whole section. Marks will be awarded from within the range of marks for that level. There is no weighting of the criteria — they are equal.

What you should do for each of these outcomes

These outcomes all concern the final stages of the investigation whereby you independently contextualise, analyse and summarise your findings and data, and draw conclusions.

Conclusions should:

- provide a summary of all the major findings made at different stages throughout the individual investigation
- give a synthesis of the relevant geographical links that have been uncovered and how they relate to the broader geographical content
- consider the evidence presented in the investigation, drawing out geographic implications and developing a series of summary ideas
- relate findings back to the original aims and focus of the investigation; conclusions may thus be the place to introduce a definitive answer

You should recognise that conclusions are often ambiguous and even unclear (and this should always be acknowledged as part of the write-up using words such as 'partial', 'tentative' or 'inconclusive'). This is not an admission of failure. Most often it arises because geographical environments and processes are complex, and data collection cannot be managed to create a reliable outcome because of the range of other external factors influencing the results.

As both AO1 and AO2 feature in the mark scheme here, you should apply your existing knowledge, theory and concepts to order and understand the field observations and identify their relation to a wider context. You should write up the field results clearly, logically and coherently using a range of presentation methods and extended writing. You should demonstrate the ability to answer a specific geographical question (or hypothesis) and do this by drawing effectively on evidence and theory to make a well-argued case. The final stages of the investigation require evaluation and critical reflection on the investigation. It is important that you evaluate the findings of your investigation and reach a balanced and supported conclusion on your own.

Critical reflection

Critical reflection is an opportunity to look back over *all* the fieldwork and research and to identify any shortcomings. It is also the chance to think more widely about the meaning of your results, comparing them with similar studies, noting and explaining similarities and differences.

It is almost certain that all NEA investigations will encounter some limitations of methodology. The most obvious concerns the choice and delivery of the sampling strategy and, as a result, the amount and quality of data collected. In questionnaire surveys, for instance, obtaining high-quality and representative samples can be difficult. This is because this type of survey work should be based on a stratified approach using the small output area statistics for a local area. Another problem is that rejection rates for street interviews are high and students may resort to interviewing anyone willing to respond. There is also the problem of visitors (as opposed to residents) — a profile of them is a likely 'unknown' so it becomes difficult to create a reliable sampling design.

Any deviation from the 'best' sampling strategy will create potential for lower reliability and, ultimately, affect the validity of the investigation. Sometimes the investigation inadvertently introduces bias. Insufficient sample data is another problem for many investigations. Insufficient data and information (primary and/or secondary) make statistical analysis difficult which, in turn, provides little confidence in the reliability of results. This problem can be tackled at the planning stage by identifying the statistical test to be used for analysis and the minimum sample size needed to obtain statistically significant results. Where there is doubt it is always best to collect a larger rather than a smaller sample. Therefore it is often useful to critically reflect in terms of:

- errors linked to sampling and design
- problems that are operator related (e.g. mishandling equipment)
- those which are directly linked to equipment (e.g. level of accuracy and resolution)

Your evaluation section should also suggest improvements and make recommendations that would improve the overall reliability of the investigation.

The 'big' picture

You should also place your study in a wider geographical context as well as comment on how it helped extend your understanding. In order to achieve these objectives, it is necessary to revisit and reconsider the literature/research elements of the investigation and then to make other links, some of which might be far-reaching.

As a **summary**, the following steps and questions will help you to conclude and evaluate your fieldwork investigation:

- look again at your initial aim or aims, commenting on the suitability of your chosen location for what you wanted to do
- review and evaluate your methods for collecting both primary and secondary data, pointing to strengths and weaknesses (or limitations)
- review and evaluate your choice of research question(s) or hypothesis(es), discussing their appropriateness in the light of what you accomplished
- develop your analysis into broader conclusions linked to geographical theory and/or what you found in your particular location
- were your conclusions to be expected, or was there something about your locality which threw up unexpected or unusual results?
- would you do things differently if you were to start again?
- suggest other avenues of enquiry that may have arisen

Writing up the report

Another important feature is the final write-up of the report. The specification states:

> **It is recommended that students write between 3,000 and 4,000 words for their independent investigation. Students will not be specifically penalised on the basis of the length of their written report. However, excessively high or low word counts may restrict students' ability to evidence the skills outlined in the marking criteria.**

Keep all your investigation work together in a separate folder. Organise this into sections for easy retrieval. The final write up of your investigation should be well structured, logically organised, and clearly and concisely written. There are three aspects of this process that you should consider: structure, language and presentation.

Exam tip

Your teacher/field study tutor must not:

- provide templates or model answers
- provide specific guidance on errors and omissions which limits your opportunities to show initiative yourself
- mark work provisionally and share that mark so that you may then improve it
- give specific guidance on how to make improvements

(a) Structure

The structure of the report should help the reader to understand it, and should also assist you in organising it logically. The following checklist provides a generalised structure to your report.

- Candidate 'Geography independent investigation' form
- Title page and contents page
- Executive summary (not a requirement, but advisory)
- The introduction — aims, research questions/hypotheses/issues being examined and scene setting
- Sources of information used
- Methods of data collection
- Data presentation, analysis and interpretation
- Conclusion, including overall evaluation
- Appendices and bibliography

You do not need to write these in the order given. Indeed, it may be easier if you do not. For example, the executive summary (if you decide to write one) is perhaps best written at the end of the process, as it is only at this stage that the 'whole picture' can be described. The following is a suggested order of completion.

1. **Data presentation, analysis and interpretation**

   This section is where you present and analyse your findings. At this stage you will have collected the data, sorted them and selected the most useful. You will know what you have found out and what it all means. Your results will be complete and they will be most fresh in your mind at this time. You should be able to interpret each separate section of your results and formulate conclusions for each one. The 'whole picture' may begin to appear in your head.

2. **Sources of information and methods of data collection**

   Now you can write about what information you collected and the methods you used. Do not forget to discuss any limitations of your methods of collection, or also of the data sources themselves.

3. **The conclusion**

   This should include a summary and an evaluation of all the major findings of your investigation. Do not present anything new to the reader at this stage. Towards the end of this section, try to draw together the sub-conclusions from each section of the data analysis into one overall conclusion — the 'whole picture'.

4. **The introduction (aims, research questions/hypotheses/issues and scene setting)**

   Having written up the bulk of the enquiry, you can now write the introduction, making sure it ties in with what follows. This section is intended to acquaint the reader with the purpose of the enquiry and the background to it.

5. **Appendices and bibliography**

   The appendices comprise additional pieces of evidence that may be of interest to the reader, but are not essential to the main findings. The bibliography provides detail of the secondary sources that have been used in your research, either as guides or as sources of information. Remember that any diagrams or text you have used or copied from secondary sources must be acknowledged — see p. 57.

6 Contents page

All sections of the report should be listed in sequence, with accurate page references.

7 Title page

This states the title of your report. Include also your name, candidate number, centre number and date of completion.

8 Executive summary (not a requirement, but advisory)

An executive summary should provide a brief statement (no more than 250 words) covering all the main aspects of the investigation. A good executive summary introduces the subject of the full report, refers to its aims and provides a brief synopsis of the findings. A very good executive report will tempt the reader into reading more by being comprehensible, interesting and stimulating. It should also make sense and read as a separate document from the full report.

(b) Language

The quality of language that you use in writing up your investigation is important. You are entirely in control of this aspect of the process and your style of writing must be appropriate for this exercise. You should avoid poor or inaccurate use of English language. In particular:

- your sentences should be grammatically correct and well punctuated
- your writing should be well-structured, with good use of paragraphs
- your spelling must be accurate (use a dictionary or your PC spellcheck)
- you must be clear in your use of specialist terminology and in the expression of your ideas
- you should be aware that the assessment of your work may be influenced by the above aspects of your writing

Proof reading is an important part of this process. Prior to submission, make sure you read through the draft from start to finish and mark any places where there are errors or inconsistencies. If possible, ask someone to do this for you — parents or relatives may help. However, it must be someone who will be highly critical of what you have written. A report littered with spelling mistakes or grammatical errors does not impress and it is possible that your PC/Mac spellcheck may miss something (for example, you may have wished to type 'than' but typed 'that' — spellcheck will recognise this as correct, whereas a proofreader, correctly, would not).

(c) Presentation

It is a fact of life that most people are influenced by presentation, and that includes teachers and NEA moderators. Bear in mind the following:

- a neatly presented handwritten or typed/word processed report will create a favourable impression, before its contents are read
- adequate heading and numbering of pages, with carefully produced illustrations, will make it easier for the reader to understand what is contained within the report

- layout is important. Do not crowd the pages with dense text, which looks unattractive. Provide adequate margins, use either double or 1.5 line spacing if using a word processing package, and make use of clear heading levels with short paragraphs. More sophisticated reports will number the paragraphs in sequence, though this is not a requirement
- it is essential to make sure that maps and diagrams are inserted in the correct place in the report — it is irritating to have to flick backwards and forwards when trying to read the document. Make sure all the references in the text are included in the bibliography
- reports written within the stated word limit tend to be better planned, structured and executed
- make sure you allow enough time to add the finishing touches that give your work the 'final polish'. It goes without saying that this time will be available providing you have not left it too near to the final deadline

You should now be in a position to submit your finished product, confident in the belief that it is the best you could have done.

Exam tip

Make sure you get a signed receipt from your teacher that he/she has received your report. Unfortunately reports can go missing occasionally, and you may need this as insurance.

Sample themes for fieldwork

Landscape systems, processes and change

The study of landscape systems is traditionally well suited to fieldwork. A key field skill for physical geographers is observation, asking questions such as 'why is this landscape like that?'. The ability to observe landforms in the field, to systematically record those observations and then apply classroom knowledge of the environment and processes to explain the formation of the landforms observed is crucial. Producing annotated field sketches is a good way to formalise this process; annotating photographs in the field using appropriate apps is another option.

The observation element within fieldwork is also a great opportunity to collect data, whether these are till fabric data from drumlins, gravel size data from coastal spits, or sand transport data from sand traps on coastal dunes. Some more specific ideas are listed below.

Glaciated landscapes

- Investigation of the size (width, height of back wall) shape, orientation and distribution of corries in a defined area.
- Investigation of the distribution and characteristic features of a glaciated valley (long and cross-sections, occurrence of striations, distribution of erosional and depositional features, postglacial modifications).
- Investigation of the distribution and formation of depositional features (glacial versus fluvio-glacial deposit analysis — size, shape, stratification) in an area of lowland ice sheet glaciation.
- Investigation of the size (height/width/length), distribution, shape and stoss end orientation of a drumlin or series of drumlins.
- Investigation of scree to measure slope, degree of sorting, mapping of source and extent of scree and vegetation colonisation to assess if scree is an active or fossil feature.

- Investigation of glacial till: till fabric analysis (situation, orientation, size and shape) to map provenance and movement of ice in a defined area.
- Investigation of kettle holes/lakes to investigate plant succession (i.e. a hydrosere study).
- Investigation of vegetation succession on moraines (i.e. a lithosere study).
- Investigation of discharge from meltwater streams in a currently glaciated environment.
- A survey of glacier mass balance in a currently glaciated environment.

Coastal landscapes

- Investigation of wave characteristics (wave height, frequency, wavelength) along a stretch of coast.
- Investigation of changing erosion and deposition on a stretch of coast before and after a storm to look at the impact of processes on coastal features (possibly using previous fieldwork records).
- Investigation of raised beaches to look at their distribution, height and postglacial modifications.
- Investigation of coastal erosion features: cliff height and profiles, mapping of incidence of faults, joints and bedding planes to study the distribution of micro features, e.g. caves, arches and stacks; the relationship between erosional features and geology; vulnerability of cliffs to collapse.
- Investigation of beach profiles: long and cross transects to map changes in beach material, gradient, pebble length and pebble roundness along a transect from low to high tide and across the width of the beach.
- Investigation of a spit using a range of transects to study shape, size and type of deposits on windward and lee sides.
- Investigation of sand dunes using transects to show dune topography, plant zonation and succession, studying changes in physical features (infiltration, pH, wind speed, percent of bare ground) and/or associated changes in biotic characteristics (percent plant cover, species diversity, plant height) (i.e. a psammosere study).
- Investigation of a salt marsh using transects to show salt marsh topography, plant zonation and succession, studying changes in physical features (soil type, pH) and/or associated changes in biotic characteristics (percent plant cover, species diversity, plant height) (i.e. a halosere study).
- Investigation of impact of humans on coastal environments — footpath erosion, outdoor recreation (such as moto-scrambling, mountain biking, pony trekking), trampling of dunes, beach litter.
- Investigation of a coastal management scheme(s) along a stretch of coast threatened by either erosion (natural or human) or flooding.
- Investigation of the impact of management structures on sediment transfer, e.g. groynes.
- A cost-benefit analysis to study the effectiveness of shoreline management plans.

Tectonic processes and hazards

- Investigation of perceptions of the characteristics and (likely) impacts of a hazard (earthquake, volcano, drought, storm).

Shaping places

Within this section of the specification you will study at least two places: the place where you live or study, and one or more contrasting locations. This approach lends itself to both quantitative and qualitative methods of study, and also to fieldwork and individual study. In contrast with the other human geography core themes — Globalisation and human systems and Geopolitics — this theme should be accessible to all students seeking to conduct independent fieldwork. Your local environment provides much scope for fieldwork, which can begin with a range of local studies and explorations. You might conduct your fieldwork by walking down local streets where you may observe and record the different local and global connections, and how these are represented and meanings made from them, in your local areas.

Some examples of contexts through which you may do this include investigations of 'clone towns', regeneration, how or why some places are rapidly changing compared with others, the varied and changing demographics or social make up of a place, and connections within and between places. The ways in which different groups of people may experience and perceive places differently may provide ideas for fieldwork. Examples include the distinctive ways in which people with disabilities experience places and are sometimes excluded from them, whether through the architecture of the place itself or the codes of behaviour and attitudes that prevail there. Local surveys could include how residents understand and see the places in which they live, and how their understandings may sometimes contrast with governmental and corporate representations, such as those in place marketing or planning documents.

Some of these ideas could be represented by these questions, all with a focus on fieldwork.

- What is unique/special about your place? You could take photographs, make sketches about particular elements of the built environment that help you identify with the place. You might want to talk to people and find out what makes it special to them.
- Who are the people using the spaces? Consider age, gender, religion, etc. What is their spatial distribution? This is a study of ethnography. Try to map and record the characteristics and variations.
- Who is marginalised or excluded from this place? Who is not well represented/found here, e.g. young, old, disabled. Is there a visible 'underclass' (this may include people begging on the street etc.)?
- What does the design and architecture say about the place? Consider what the age and style of buildings tells us about the character of a place, and how that influences our opinions, especially as visitors. Also important is an idea of scale and use of different types of building materials, e.g. modern (glass, steel, polished stone) versus traditional (rough stone, brick, concrete).
- What is the 'feel' of the neighbourhood? Use adjectives to describe what your instincts, and those of others, are about particular places; create a map to show this.
- How could this place be improved? What parts of the urban environment are most disappointing (possibly including neglect, dirt, dereliction)? Again, map and provide evidence to support your ideas.

- How is this place different from another area/your other study area? Carry out a comparison of differences: focus on the 'feel', e.g. streetscape, furniture, design, architecture and, of course, the people.
- How is this place's representation shown in various forms of media, such as advertising imagery, poem, song, art and even soundscapes?

Themes for Regenerating places and Diverse places

- For any urban or rural regeneration/re-branding/re-imaging initiative/project: an assessment of its success to assess environmental, economic, social and cultural impacts.
- For any regeneration/rebranding/re-imaging initiative/project: an assessment of its level of sustainability in terms of linkage and involvement to local community, conflicts, economic success, quality of jobs, impact on poor people in an area and likelihood of being value for money and a permanent success.
- Research into how regeneration affects a local environment; 'before and after' investigations.
- Examination of the role of business-led initiatives, community-led schemes, pop-up shops and any changes driven by consumers in an area (such as gyms and click-and-collect points).
- Investigation of architectural reminders of the past; linked to an investigation of the physical growth of a village/town; a place-scape or city-scape investigation.
- Changes in functions and characteristics of an area (urban or rural); economic changes in an area; the impact of these on identity.
- Changes caused by major planning schemes in an area (urban or rural) such as redeveloped town centres, housing estates, gated communities, science parks, bypasses; the impact of these on identity.
- Investigation into environmental changes linked to specific people — e.g. the disabled, the elderly, the young; the impact of these on identity.
- Investigation of socio-economic inequalities and/or variations (such as education, health, income, life expectancy, ethnicity) within an urban area, spatially and across societies; possible link to cultural diversity.
- Investigation of variations in levels of deprivation in urban areas: environmental quality, unemployment rates, crime levels, housing tenure, council tax bands, benefit uptake.
- Investigation of the environmental quality of various parts of a place.
- Investigation of the environmental, social and economic impacts of a single, large tertiary employer, e.g. a hospital complex or a sports stadium.
- Investigation of impact of tourism on honeypot sites.
- Investigation into the health profile of a place, a district or a community in that place.
- Investigation into the health provision within a town, district or community.
- Investigation into food supplies within an area, possibly involving the concept of 'food deserts' — areas that have difficult access to supermarkets and/or fresh fruit and vegetables, with an associated high incidence of fast-food outlets.
- Investigation into any aspect of leisure provision in an area, and the environmental influence of/impact on that provision; possibly linked to extent of open space in an urban area.

- Investigation of changing service provision in villages.
- Investigation of changes in, or characteristics of, suburbanised villages: population size and structure, employment characteristics, housing and community spirit.
- Investigation of changes in rural areas associated with change in the rural economy: holiday homes, population size and structure, employment and house prices and problems of service provision.
- Investigation of changes in building age, type and quality resulting from gentrification.
- Investigation of employment changes (quality and number of jobs) in redeveloped areas.
- Investigation of student districts in urban areas: population characteristics, service provision, attitudes of local residents and housing quality/tenure.

Other themes based in urban areas

- Investigation of urban microclimate — measuring temperature, relative humidity, precipitation, wind strength, light intensity along a transect from the inner-city to the suburbs, possibly also recording building height and land use changes.
- Investigation into the actual and/or perceived impact of climate change on an area.
- Investigation of impact of urban development on hydrology within a named place.
- Investigation of a Sustainable urban Drainage System (SuDS) scheme — rationale, strategies and success.
- Investigation of an urban river restoration scheme — aims, attitudes, success.
- Investigation of atmospheric pollution in urban areas — nature, causes, impacts, management and success.
- Investigation of sustainability at/in a variety of scales/formats within an identified urban area.

Globalisation and human systems and Geopolitics

These themes might seem relatively inaccessible to students seeking to conduct independent fieldwork. However, some themes may overlap with Shaping places, particularly when looking at the impact of globalisation and global governance on a local scale.

Here are some ideas.

- Investigation of the impact of migration on a particular community: provision of shops, services, schools, places of worship, distribution of ethnic groups, official services (e.g. language related); indices of segregation.
- Investigation of the distribution of ethnic food outlets and restaurants in a designated area.
- Investigation of the presence of global brands (retail, industrial) in an area.
- Investigation of global tourism trends (origins or destinations) within a place.
- Investigation of accessibility and/or connectivity of a place to/from the rest of the world.
- Investigation of global influences on the characteristics and lives of people in a place.
- Investigation of how people use social networks to maintain contact with families.
- Investigation of the impact of Chinese or Indian or Pakistani diaspora in a named area.
- Investigation of the impact of FDI in a named place, e.g. Tata in South Wales/ Scunthorpe, Chinese-funded development around Manchester airport, Hinkley Point or Liverpool city centre.

- Investigation of a beach to look at distribution and type of sea borne materials (after a storm and post clean-up) as well as land supplied litter and waste.
- Investigation of water quality and management of water quality in coastal areas (blue flag beaches) — linked to management of the seas.

Physical systems and sustainability

The water cycle and water insecurity

Several themes can arise here. **Infiltration** is a key process within a river catchment. Infiltration rates may be affected by a range of factors such as surface vegetation cover, antecedent weather conditions, soil moisture, soil texture, slope and soil compaction, allowing you to conduct a range of related but distinct investigations in a constrained area.

Catchment discharge is a fundamental parameter in the drainage basin **water balance**. Use of secondary rainfall and runoff data will allow you to construct water balances for catchments. Measuring stream discharge and rainfall in the field could help you to understand the potential issues associated with these measures. Measurement of rainfall in multiple simple rain gauges will allow you to examine spatial variation in rainfall and its potential impact. Evaporation rates can be measured in the field, and evapotranspiration rates can be obtained as secondary data from the Met Office.

Estimating **fluxes of materials** in rivers involves the measurement of discharge and of the concentration of the materials, such as sediment. Flux is calculated as the product of discharge and concentration. For example, you might measure river discharge either from a site with known stage-discharge relationships (e.g. a weir) or by measurement of velocity and river cross-section. Sediment concentration could be measured by filtering water samples.

Further fieldwork ideas include:

- Investigation of throughfall and how it varies through the year, and by vegetation type.
- Investigation of drainage basin characteristics: land use, vegetation, slope, soil permeability/infiltration and their impact on river discharge.
- Investigation to compare the characteristics of two drainage basins/catchments.
- Investigation of river discharge over selected times in a year to look at river regimes in relation to season.
- Investigation of a storm event and its impact on discharge in a small stream catchment.
- Investigation of flooding recurrence levels and areas of flood risk/vulnerability.
- Investigation of the impact of a sustained period of drought on water supply and water use, vegetation, sales of summer products (ice creams, salads) and summer activities.
- Phenological investigations to look at the impact of climate change on natural and human activities (e.g. appearance of catkins or snowdrops, first and last marking of lambs, putting sheep inside, lambing).
- Investigation of the impact of human activity (e.g. urbanisation, agriculture and deforestation/afforestation) in drainage basins.

- Investigation of strategies to increase water supply in an area.
- Investigation of strategies to manage water consumption in an area.
- Investigation of strategies to manage water quality in an area.
- Investigation of a conflict over water at a local scale.

The carbon cycle and energy security

Several themes can arise here. The **stock of carbon** within woodland can be estimated. There are standard equations to estimate living biomass of trees from the diameter of the tree measured at 1.3 m height. Tree biomass is 50% carbon so it is a simple conversion to work out how much carbon is stored in the tree. Where tree age can also be estimated, either from the girth of the tree, knowledge of the site or from tree-ring evidence from similar felled trees, then the rate of carbon sequestration as a mass of carbon per year can be calculated. Hence, you could estimate both the stock of carbon and the flux within a woodland context.

Estimates of **changes in carbon stocks** in peatlands can be made. Peatland depth can be measured by probing. Multiple probings of depth in an area can be used to estimate peat volume at a site. Peat volume can be converted to organic carbon stocks by knowing typical peat densities (around 0.1–0.2 g C per cm^3). Where the age of local peatlands is known, the total carbon stock can be converted to an average flux over this time period by dividing the stock by the age to give flux in units of grammes of carbon per metre squared per year.

Estimating **fluxes of carbon materials** in rivers involves the measurement of discharge and of the concentration of carbon. Flux is calculated as the product of discharge and concentration. The organic component of sediment can be estimated as the fraction lost after an hour in a furnace at 550°C. Carbon content of sediment is typically 50% of organic content. The fluvial particulate carbon concentration (the amount of carbon being transported in the sediment) in grammes per cubic metre multiplied by discharge in cubic metres per second will give a carbon flux in grammes per second. The same calculation can be applied to dissolved carbon concentrations estimated by colourimetry. This approach is particularly relevant to peatland streams where 'brown water' is indicative of high dissolved carbon concentrations.

Further fieldwork ideas include:

- Investigation of the impact of a thermal power station (oil, gas or coal fired) on local microclimate, water air pollution levels, transport movements and employment.
- Investigation of the social, environmental and economic impact of a nuclear power station on a designated area.
- Investigation into the patterns of energy consumption in two contrasting communities.
- Investigation of the impact of energy efficiency measures on a community, to include issues such as recycling, use of solar panels, etc.
- Investigation of the impact/potential impact of a solar energy farm on a place.
- Investigation of potential sites for the location of wind farms and/or the impact of existing wind farms.
- Investigation of loss of coal mining in a former mining area, exploring impact on image, economy, culture, health issues and environment, and measures to rebrand.
- Investigation of a conflict over energy at a local scale.

Summary

After studying this section, you should:

- be aware of the specification requirements of the fieldwork elements for both the AS and A-level assessment
- understand the nature of the fieldwork that should be undertaken prior to the assessment of that fieldwork within the AS examination
- understand the demands of the mark scheme for the NEA at A-level, and be aware of what you should do to satisfy those demands
- understand how to structure, conduct and write up the individual investigation that constitutes the NEA
- be aware of the wide range of potential themes and ideas that can be used to formulate the title of your own individual investigation for the NEA

Synoptic skills

What are synoptic skills?

Synoptic skills involve your ability to 'think like a geographer' and 'see the big picture'. Most exam questions test your knowledge and understanding of one topic only. Synoptic exam questions test your ability to:

- link one topic to others
- make connections between different places
- show that some topics, such as global climate change or globalisation, seem to have an impact on almost all areas of geography

It is important not to see the topics you study as separate, but rather as linked.

Synoptic themes

There are three synoptic themes that appear in all topics of your course:

1. players
2. action and attitudes
3. futures and uncertainties

It is important to understand these themes, and how they link together (Figure 29).

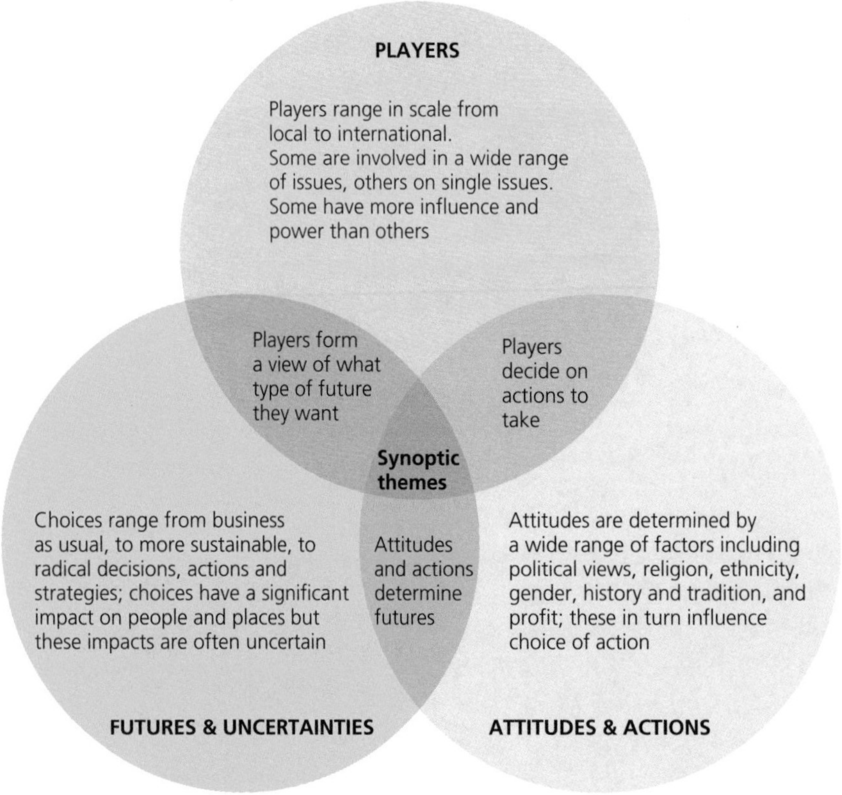

PLAYERS

Players range in scale from local to international. Some are involved in a wide range of issues, others on single issues. Some have more influence and power than others

Players form a view of what type of future they want

Players decide on actions to take

Synoptic themes

Choices range from business as usual, to more sustainable, to radical decisions, actions and strategies; choices have a significant impact on people and places but these impacts are often uncertain

Attitudes and actions determine futures

Attitudes are determined by a wide range of factors including political views, religion, ethnicity, gender, history and tradition, and profit; these in turn influence choice of action

FUTURES & UNCERTAINTIES

ATTITUDES & ACTIONS

Figure 29 Synoptic themes

Exam tip

Learn the three synoptic themes and be prepared to refer to them in the exam, even if you are not directly asked to.

These synoptic themes could be referred to directly in AS or A-level exam questions. It is more likely that you will need to refer to them within your answers, even when you have not been directly asked to.

The synoptic themes can be considered in the context of all of the topics you have studied. An example of this is shown in Figure 30 but this way of thinking should be applied to all topics.

Players
- Are TNCs, governments or IGOs most responsible for driving the process of globalisation?
- Why are the impacts of globalisation so varied between different types of worker, migrant, age groups and genders?

Actions and attitudes
- Why are some players such as NGOs, environmental groups, indigenous cultures and some political organisations against globalisation?
- What actions have been taken by players to promote globalisation, and also to restrict it or lessen its negative consequences?

Futures and uncertainties
- In the long term, will globalisation reduce or increase inequality on local to global scales?
- Can globalisation be sustained by available natural resources and/or will the environmental consequences of globalisation eventually become too great?

Figure 30 Synoptic themes and globalisation

AS synoptic skills

In the AS exams, there are synoptic questions on both Paper 1 and Paper 2. They require you to make links between compulsory and optional topics, as shown in Table 23.

Table 23 AS synoptic assessment

AS exam paper	Question number (all 16 marks)	Compulsory topic	Optional topic	Examples of links between topics
Paper 1 (Physical)	4	Topic 1: Tectonic processes and hazards	2A Glaciated landscapes and change	Impacts of volcanic activity on glaciers and ice sheets (lahars, jökulhlaups) Sea-level change caused by tectonic processes and their impact on cold environments
	7		2B Coastal landscapes and change	Impact of tsunami on coastlines How volcanic activity can create new coastal landscapes
Paper 2 (Human)	4	Topic 3: Globalisation	4A Regenerating places	The role of TNCs and foreign investment in urban regeneration The impact of the global shift on UK industry and inner cities
	7		4B Diverse places	Impact of economic migration on the cultural diversity of cities Impact of EU immigration on rural culture in the UK

The exam context

Each AS synoptic exam question is a data stimulus question. You will be asked to study a context and a number of different resources. The context will:

- be about a real place, usually a small location such as a stretch of coastline, part of a city or a rural area
- have some text you will need to read
- have other resources such as photographs, data tables, graphs, satellite imagery, GIS data

The whole context will be no more than two pages long.

Your task is to understand the resources you have been given (comprehension) and use them to answer the synoptic question.

Exam tip

Remember which two topics your question links together and be careful not to drift into other, unrelated topics.

Key AS exam skills

When answering the AS synoptic questions you need to use the information and data provided in the context as fully as possible.

1. Refer directly to all of the resources in the context, including any images and text.
2. Quote data in your answer, especially numerical data from graphs and tables if these are included.
3. Numerical data should be analysed: look for trends or anomalies, calculate means and ranges.
4. Use all of the information you have been given as evidence to make a case.

You also need to be synoptic.

1. Stress links and connections between the two topics in terms of cause and effect.
2. Use examples you have learned that are similar to the context, or that provide a different perspective.
3. Use key terminology from both topics.
4. Use any key concepts, models or ideas that are relevant.

The highest marks will only be achieved if you go beyond the information and data provided in the context and add in some additional material from your wider course.

Exam tip

Remember to refer to all of the resources you have been asked to look at, and quote data and information in your answer.

A-level synoptic skills

A-level synoptic skills are tested in Paper 3, a 2-hour 15-minute exam marked out of 70 and worth 20% of the A-level qualification. This exam is based on a resource booklet of information, likely to be about 8–10 pages long. It will:

- be set in a real place and consider a complex geographical issue
- contain information and data in the form of text, maps, data tables, graphs, images and the views of key players
- contain some complex information, such as scatter graphs, detailed GIS maps and large tables of data

Questions on Paper 3 will be in a sequence and become more demanding as you move through the exam paper:

- 4–6 mark questions focused on explaining single processes and themes

- 4–8 mark questions that could include calculating and interpreting statistical tests, and analysing numerical data and relationships
- 18 and 24 mark questions require you to evaluate issues, drawing on data and information from several different sources in the resource booklet and using your wider knowledge and understanding

It is not possible to get maximum marks on the 18 and 24 mark questions by *only* using the information in the resource booklet. You will have to use your wider knowledge and understanding as well.

Exam tip

You will need to spend 5–10 minutes at the start of the Paper 3 exam reading the whole resource booklet to get an overview of the tasks you will have to complete.

Being synoptic

There are several different ways in which you can be synoptic in the exam, and demonstrate wider knowledge and understanding and the ability to make connections between topics. These are shown in Table 24 and some are explored in more depth below.

Table 24 Being synoptic in the Paper 3 exam

Parallel examples/case studies	Look at the resource booklet context you have been given ■ Do you know of any similar and/or contrasting examples or case studies? ■ Could you use evidence from these in your answer? This could be evidence in support of a point or argument, or for a counter-argument
Theories and models	You have learned a number of models and theories such as: ■ the environmental Kuznets curve ■ capitalism ■ pressure and release model (PAR), Park's model ■ the hydrological and carbon cycles ■ Clark-Fisher model ■ modernisation and dependency theory You may be able to refer to one or more of these to put the resource booklet issues in a broader context
Global themes	There are several global themes which link to almost any geographical issue, because they affect most corners of the world. These include: ■ global warming ■ population growth and urbanisation ■ resource demand growth and biodiversity decline ■ globalisation Linking to one or more of these themes can show you are capable of making connections from the small-scale context in the Paper 3 resource booklet to wider global issues
Contemporary events	Recent news events can be useful, as making reference to them shows up-to-date awareness
Specialist concepts	There are a number of specialist geographical concepts in the specification: causality, systems, feedback, inequality, identity, globalisation, interdependence, mitigation and adaptation, sustainability, risk, resilience and thresholds Reference to any of these will show your wider understanding

Exam tip

Most synoptic references you make should be no more than a sentence or two of additional information, not long paragraphs about case studies you have revised.

Smart example use

A skill worth learning is the ability to take an example or case study learned in one topic and apply it to a different topic.

This involves a little bit of lateral thinking, but is not hard to do. It makes your case studies more flexible so you can use them as evidence in a wide range of questions. A good example is the Three Gorges Dam in China. You will probably have encountered the Three Gorges Dam in Topic 5, The water cycle and water insecurity, as an example of a large-scale 'techno-fix' solution to water supply. However, as Figure 31 shows, the same example can be used in a number of different contexts.

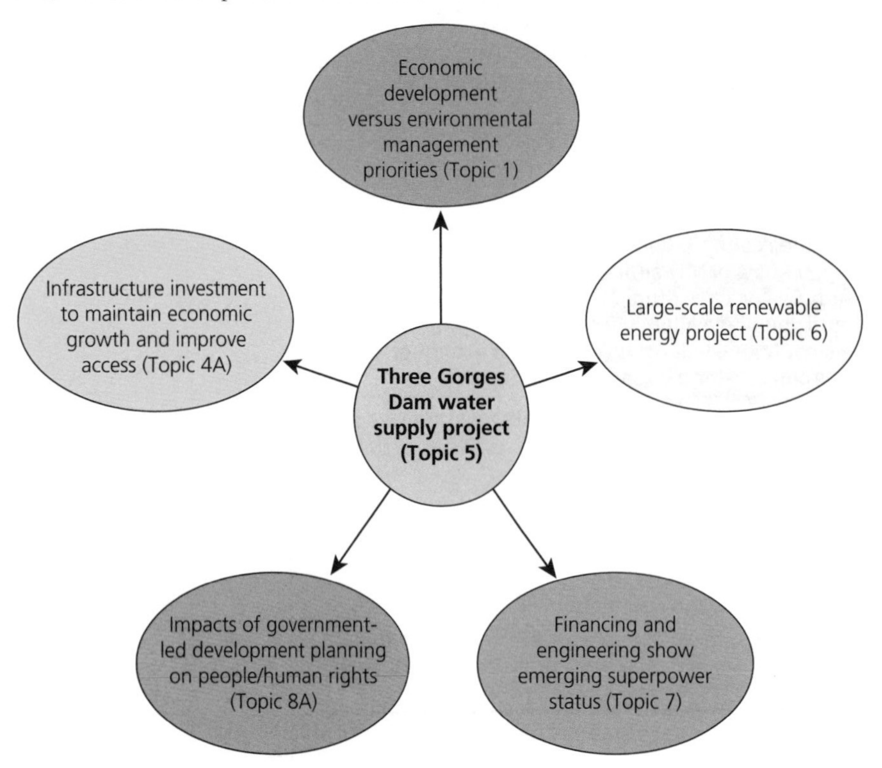

Figure 31 Smart example use

To use examples and case studies in this smart way, you must be selective and show application:

- choose which information (select) to use (apply) to the question
- do not fall into the 'tell the examiner everything I know about ...' trap

The evidence you use needs to be relevant to the question.

Exam tip

During revision, think about how you could use a case study or example from one topic in another topic.

Using criteria to make judgements

At A-level, longer synoptic exam questions will use the following command words:

- **assess**, meaning using evidence to determine the relative significance of something, giving balanced consideration to all factors and identifying which are the most important
- **evaluate**, meaning to measure the value or success of something and provide a balanced and substantiated judgement/conclusion

Both of these command words involve making judgements. Some exam questions provide a structure for this by including phrases such as:

- costs and benefits
- advantages and disadvantages
- strengths and weaknesses

These phrases make it clear that you need to consider both sides of an argument before coming to a conclusion.

In order to make detailed judgements you need to go a little further and decide on some criteria to measure any impacts or consequences against. These can include:

- **temporal criteria:** short-term versus long-term — for instance, new coal-fired power stations might have short-term benefits in terms of cheap, reliable electricity supply but long-term costs in terms of carbon emissions and localised pollution
- **spatial criteria:** small- versus large-scale — for instance a new HEP power station might be beneficial in terms of national energy security, but have costs for local people who are displaced by its construction
- **impact criteria:** economic, socio-cultural, environmental and political — these criteria consider different aspects of geography and how they are affected by a process, change or decision (Table 25)

Table 25 Impact criteria

Economic	Socio-cultural	Environmental	Political
Impacts concerning economic development, jobs, incomes, businesses and taxes	Impacts on people and their health, housing, diet, education and quality of life	Impacts on ecosystems and physical systems and cycles	Impacts on decision making, decision makers, planners and governance

At A-level, you should not expect questions to include criteria as, for example, this one does:

- Evaluate the economic, environmental and political costs and benefits of globalisation for different players.

It is more likely that you will have to decide on the criteria yourself:

- Evaluate the costs and benefits of globalisation for different players.

Exam tip

You should plan 18 and 24 mark questions in Paper 3 before starting to write your answer; give yourself some thinking time to make sure you have fully understood the question.

Sustainability

The theme of sustainability runs throughout the specification. It is essential that you have a good understanding of it. This is because sustainability can be used as your criterion for making judgements.

Sustainability has two meanings. Originally it was used in the context of human development, i.e. improvement in people's lives or social and economic progress. The definition 'sustainable development is development that meets the needs of the present without compromising the ability of future generations to meet their own needs' dates from the 1987 UN World Development Report.

This definition is often shown as consisting of three pillars (Figure 32). For something to meet the criteria for sustainable development, it should:

- promote economic development, but this should be equitable (fair) and not lead to greater inequality
- promote human development (health, education, gender equality, etc.)
- have minimal environmental impacts, i.e. limit pollution, use few natural resources, not damage ecosystems and physical systems

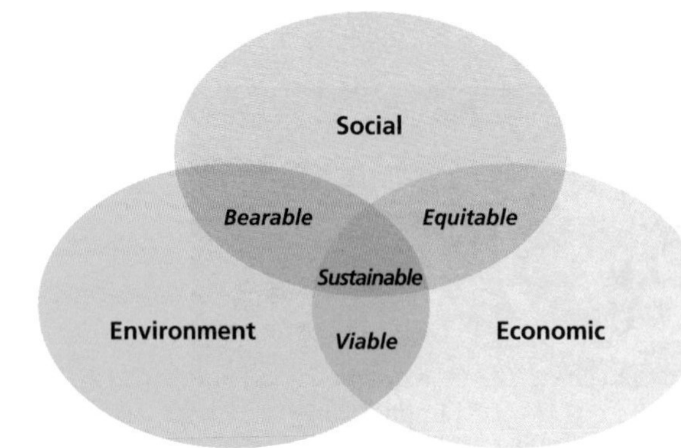

Figure 32 The three pillars of sustainable development

In Paper 3, the model in Figure 32 can be used as your criteria to make judgements.

Since the mid 1990s the words 'sustainable' and 'sustainability' have often become separated from the word 'development' present in the 1987 definition. They are used to indicate if something is 'green' or 'ecofriendly'. This is sometimes called environmental sustainability and is a narrower concept than sustainable development.

Exam tip

Sustainability will almost always be relevant to a Paper 3 exam, so you should plan to use the concept and make sure you fully understand it.

Specialist geographical concepts

The specification has a list of specialist geographical concepts (Table 26). These appear in different topics, often in more than one. It is important that you understand the meaning of each of these because:

- they could appear within a question on Paper 3, or within the resource booklet
- if you refer to them in your answers you would be being synoptic, i.e. making a link to an important concept

Table 26 Specialist geographical concepts

Concept	Meaning
Causality	Connections between cause (why) and consequence as part of a process; in many cases processes have many causal factors, i.e. are complex
Systems	Many interacting component parts, producing a complex 'whole'. The hydrological cycle and carbon cycle are both examples of systems
Feedback	Positive feedback causing further change and instability to a system, or negative feedback returning a system to equilibrium
Inequality	At all scales, differences in opportunity, access to resources or outcomes (e.g. health) between different groups
Identity	The beliefs, perceptions, characteristics that make one group of people different to another; identity is strongly related to place
Globalisation	The set of processes leading to greater international integration economically, culturally and demographically
Interdependence	Mutual reliance between groups; it is strongly linked to globalisation
Mitigation and adaptation	Alternative approaches to management: prevention (mitigation) versus reducing vulnerability (adaptation)
Sustainability	Passing the planet and its natural systems and resources on to the next generation in as good a state as we inherited it
Risk	The potential or probability of harm/losing something of value
Resilience	The ability to cope with change, e.g. resilience to global warming or a natural hazard such as an earthquake, or a human change such as an economic recession
Threshold	A tipping point in a system; a critical level beyond which change is inevitable/irreversible

Global themes

A very useful way to demonstrate synoptic thinking in Paper 3 is to link the local or regional scale issue given in the resource booklet to wider global themes.

There are a number of global themes that provide a 'context' for almost all geographical issues because they are inescapable — they impact on almost all people in almost all places. These are summarised in Table 27.

Table 27 Global themes

Global resource 'crisis'	Rising demand for land (food), water, energy resources and minerals This is partly as a result of population growth, but even more so because of increased affluence in emerging countries Some resources have a finite supply, raising questions of rising economic cost in the near future New technologies may overcome shortages and/or increases in price
'Sixth mass extinction'	Global biodiversity loss through ecosystem destruction and degradation This is linked to resource exploitation, urbanisation, pollution and global warming The rapidity of ecosystem and species loss as a result of human activity appears to exceed 'natural' extinctions present in the geological record
Global poverty and inequality	The persistence of widespread poverty in some places such as Sub-Saharan Africa and South Asia The apparent increase in global inequality (rich versus poor divide) despite economic development in many emerging countries
Global demographic trends	Increasing global population and the uncertainty about total numbers in the future: 8, 10 or 12 billion people by 2100? Increases in the global pool of migrants and the increased pace and scale of migration flows Continued urbanisation of the population, which now exceeds 50% of the world's people living in towns and cities
Technology and connections	The rapidity of technological change especially in the area of communication and connectedness (the internet) but increasingly areas such as automation and genetic research The difficult ethical questions this raises, and economic challenges
Global warming	Possible scenarios for global warming as outlined by the IPCC and the uncertainty of future global climate change and sea levels How global warming will affect ecosystems, food production and farming, cities and natural systems such as the hydrological cycle
Globalisation	The causes and consequences of continued globalisation, including the success stories and ethical questions Why some places have become globalisation's winners whereas other places still lack global connections

Exam tip

Try to link your Paper 3 answers to one or two global themes, but no more. Your answer could become very long and not well focused on the question.

Futures

The final synoptic theme in the specification is futures and uncertainties. It outlines three contrasting futures, which are:

- business as usual
- more sustainable
- radical

It is important to understand what these might mean. Figure 33 shows what they might mean for global levels of atmospheric carbon dioxide.

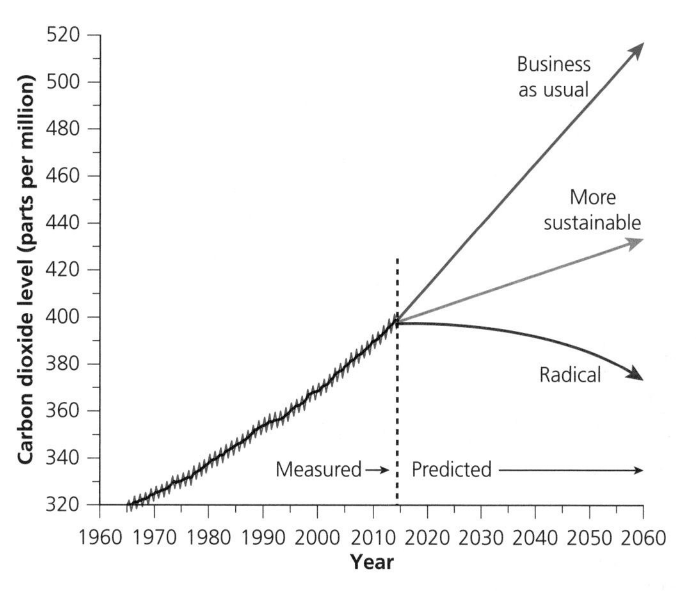

Figure 33 Future carbon dioxide levels?

- In the 'business as usual' prediction, humans continue to emit carbon dioxide as they have done since the 1960s, i.e. at an increasing rate. The implications of this are widely thought to be a warmer world, higher sea levels and widespread changes to food production, water supply and ecosystems.
- In the 'more sustainable' prediction, humans attempt to reduce the rate of emissions growth, so the changes brought by global warming happen more slowly and are not so large. However, to achieve this we would need to move away from fossil fuel use. The changes needed might not be very large.
- In the 'radical' prediction, carbon emissions need to be dramatically reduced in order to return atmospheric concentrations to levels seen in the year 2000. This would mean major changes to lifestyles, energy use and technology — but it might stop most global warming.

The same three 'futures' can be applied to many geographical issues such as:

- future global population
- the rate of biodiversity loss and deforestation
- total water use by people
- the average wealth of a global citizen

It is easy to decide which future you think might be the most desirable. However, the implications of your choice are huge. For instance, a decision to follow even the more sustainable path for total global population raises very difficult issues about how this might be achieved if it required birth rates and fertility rates to somehow be controlled. Part of being synoptic is thinking about these futures and being realistic about what might be achieved and how desirable different futures are.

Being evaluative

Perhaps the most important skill of all for Paper 3 is the ability to be evaluative. This means:

- weighing up evidence from the resource booklet, deciding what is important, and using it in your answer

- seeing both (or all) sides of an argument or debate, and discussing the merits of each
- recognising that some impacts or consequences are more important than others
- making clear judgements supported by evidence

Being evaluative involves using evaluative language. This is a set of words and phrases that show the examiner you are weighing things up before coming to a conclusion (Figure 34). Using this evaluative language is very important and you should practise using it.

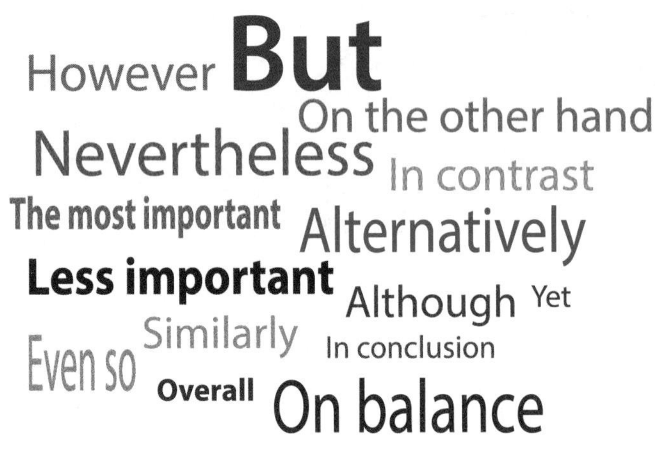

Figure 34 Evaluative language

Summary

- Being synoptic involves making links between one area of your geography course and others, to show you can make connections and think beyond a single topic area.
- This is an important skill because many areas of both physical and human geography cannot be fully understood in isolation.
- Synoptic skills are tested in both AS extended writing questions and through A-level Paper 3.
- The three synoptic themes (players, actions and attitudes, futures and uncertainties) appear in all topics and could feature in exam questions.
- You can be synoptic in the exam by using case studies and examples in a 'smart' way — perhaps not in the topic context you studied it within.
- You should also try and link to global themes and specialist concepts, which provide a context for many geographical issues.

Questions & Answers

Assessment overview

AS fieldwork

In Papers 1 and 2, the optional Sections B and C will assess fieldwork skills. Students are required to answer one question from the Physical options (Topics 2A or 2B) and one question from the Human pair (Topics 4A or 4B). It is therefore expected that you answer a fieldwork question on either Glacial landscapes and change or Coastal landscapes and change and either Regenerating places or Diverse places. Each question is worth 18 marks and is made up of two parts:

- part (a) is composed of a few short-answer questions based on resource material, worth a total of 9 marks. This material is said to be 'unfamiliar' to you
- part (b) comprises one longer question based on the skills you have developed through the fieldwork investigation you have carried out and is worth 9 marks. This material is said to be 'familiar' to you

The overall length of each paper is 1 hour 45 minutes, therefore it is recommended that you spend approximately 30 minutes on the fieldwork skills questions.

The sections below are each structured as follows:

- sample questions typical of the style and structure that you can expect to see in a fieldwork question in an AS paper
- example student answers at an upper level of performance
- examiner's commentary on each of the above

Each extended question (6 marks and over) is assessed by means of a levels-based mark scheme in the style of the examination. Study the descriptions carefully to understand the requirements necessary to achieve a high mark. You should also read the examiner's comments with the mark scheme to understand why credit has or has not been awarded. In all cases, actual marks are indicated.

Level	Descriptor
Level 1 (1–3 marks)	Shows evidence that fieldwork investigation skills used may not have been fully appropriate or effective for the investigation of the geographical questions/issue Considers the fieldwork investigation process/data/evidence, with limited relevant connections and/or judgements Argument about the investigation is simplistic and/or generic
Level 2 (4–6 marks)	Shows evidence that fieldwork investigation skills used were largely appropriate and effective for the investigation of the geographical questions/issue Critically considers the fieldwork investigation process/data/evidence in order to make some relevant connections and valid judgements Argument about the investigation may have unbalanced consideration of factors, but is mostly coherent
Level 3 (7–9 marks)	Shows evidence that fieldwork investigation skills used were appropriate and effective for the investigation of the geographical questions/issue Critically considers the fieldwork investigation process/data/evidence in order to make relevant connections and judgements that are supported by evidence Argument about the investigation includes balanced consideration of factors and is fully developed and coherent

Questions

Glacial landscapes and change

Question 1

Study Figure 1 which shows a photograph of an outwash plain to the south of a glacier in Iceland. Figure 2 shows a field sketch of the same area.

Figure 1 An outwash plain in Iceland

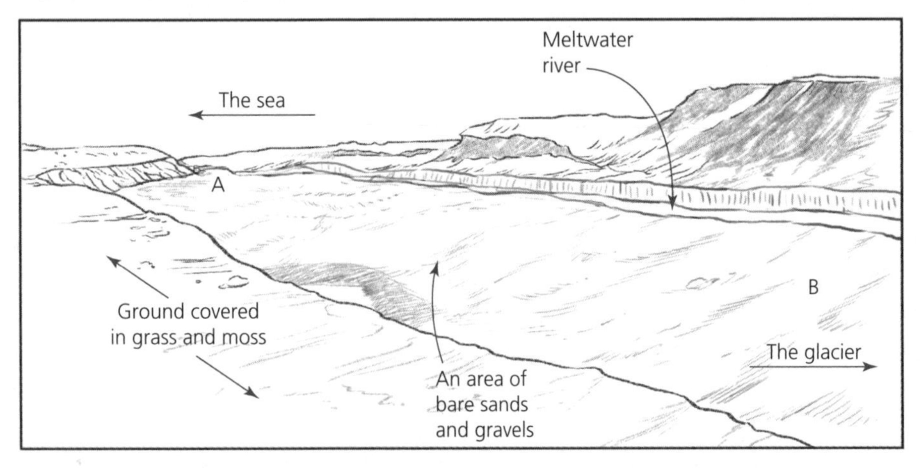

Figure 2 Field sketch of Figure 1

(a) A student collected sediment from each of the sites A and B shown in Figure 2 to investigate the differences in sediment size and roundness between them. The results of these data are shown in the table below.

Site	Mean sediment size (mm)	Average sediment roundness (*R*) — as determined by the Cailleux scale. Greater *R* values equate with greater roundness	Relative distance from the glacier
B	22.7	230	Nearest
A	17.5	568	Furthest

(i) Using the data, identify **one** difference between the sediment found at the two sites A and B. (1 mark)

ⓔ One mark for correct identification of difference. The answer must use a comparative word.

Student answer

(a) (i) From the data, it is clear to see that the size of sediment decreases from B to A.

ⓔ **1/1 mark awarded** The student makes a correct statement. They could also have said that the sediments are more rounded from B to A.

The student carried out a Student's *t*-test on the difference in the sediment size at the sites A and B.

First he stated his hypothesis: *There is a significant difference between the mean size of sediment found at sites A and B.*

His null hypothesis therefore was: *There is no significant difference between the mean size of sediment found at sites A and B.*

The outcomes of the test are shown in the table below:

Standard deviation Site B	Standard deviation Site A	Difference in means	*t*-test score	Critical value at 95% confidence level
5.1	3.6	5.2	2.6	1.73

(ii) **Which hypothesis should the student accept and why?** (2 marks)

ⓔ One mark for accepting the correct hypothesis. One mark for stating that the *t*-test score is higher than the critical value.

Student answer

(a) (ii) The correct hypothesis is that there is a significant difference between the mean values of sediment size between sites A and B. I can say this as the outcome of the *t*-test (2.6) is a larger number than the critical value of 1.73. The result is therefore statistically significant.

e 2/2 marks awarded The student identifies the correct hypothesis, and gives the right reason for this.

(a) (iii) Suggest one reason why the deposits measured at Site A are different from those measured at site B. (2 marks)

e One mark for identifying a reason why deposits at Site A and B are different and a further mark for justification of why this is the case.

Student answer

(a) (iii) The sediment is smaller at site A than B because they are further away from the glacier a and as they get washed down river in the meltwater they are eroded by bumping into each other b. Their rough edges are also smoothed b in this process and hence they become more rounded.

e 2/2 marks awarded The student provides a reason for difference a, together with two justifications b. The student has cleverly connected the two differences in the sediment.

(b) The student collected the data during a one-day field trip in summer. Explain how the student could have gone about collecting the data he needed. (4 marks)

e One mark for each statement of data collection methodology, as long as the methodology is appropriate in the context of the fieldwork activity.

Student answer

(b) Looking at the photograph, this seems quite a remote area, and so before he set out to collect the data the student should have informed someone of what he was doing and where he was going a. Once there, it is clear that some form of sampling should be undertaken. I would suggest using a quadrat — a square metal-framed object which is usually 1 m^2 a. The quadrat can be placed in a number of locations across a transect on the deposits — at both sites A and B — possibly two or three sites for each transect a. Using either random or systematic sampling within the quadrat, individual items of sediment can then be examined, their size measured and their roundness identified using the Cailleux scale a.

e 4/4 marks awarded The student makes four valid statements a.

(c) You have also carried out field research to investigate glacial landscapes and change. Evaluate the methodologies you used to present and analyse the data you collected. (9 marks)

Location of geographical investigation:

e See the mark scheme on p. 93.

Student answer

(c) Location of geographical investigation: The fieldwork I conducted was in the Nant Ffrancon area and specifically in Cwm Idwal [a].

We used the process of till fabric analysis to collect, present and analyse sediment data. We had to walk around Cwm Idwal to find an exposed moraine cut through by a stream so that it was naturally exposed and we didn't disturb or change the orientation of till by digging. When we found moraine that was exposed, one member of our group gently removed the till from the moraine and kept it as steady as possible so as not to alter the orientation. Another member of the group placed the longest side of the compass against the a-axis of the till (longest length). They then turned the outer ring so that North was in line with the needle pointing north and read the bearing. Another person then recorded this bearing [b].

To present the results of the orientation of till, I used a rose diagram (radial graph). I grouped the data into sets and shaded between the correct bearings up to the number of stones recorded. By doing this a mode could be seen [c] and so a dominant orientation would have been shown. Bearings go from 0 to 360 degrees as in a compass and so are easy to read [c].

By grouping the till results and plotting them on a rose diagram I found that it showed till with orientations different from the modal orientation. Within the till rose diagram the mode orientation was 30–60°. However, there were some other results from 210° to 240°. However, these were likely to have been due to human error on the other side of the axis. Measuring the a-axis of a pebble gives two orientations and so by using secondary data such as maps the most likely orientation will be shown 30–60° as otherwise the glacier would have been moving towards the backwall [d].

Overall my presentation method allowed me to determine the direction of flow of glaciers in this area of north Wales as they emerged from the corries [e].

e **7/9 marks awarded** The student begins by providing a clear indication of the location of the fieldwork [a]. Although the question requires an evaluation of presentation and analysis, it is fair to outline the data collection method. However, this answer perhaps provides too much detail that is not required, although it is clearly appropriate [b]. The student then moves to presentation and provides two, relatively simple, statements of evaluation [c]. The next paragraph provides a further discussion of the challenges faced in the presentation/analysis process, which recognises the importance of checking with other sources of data, in this case secondary, to explain the apparent anomalies in the radial diagram [d]. The final sentence provides a clear statement of success with a clear link between presentation and analysis [e]. The answer could have been improved with more evaluation of the technique of drawing a radial chart, and a more clear explanation as to how it is used to show orientation. The apparent anomalies do not in fact matter in this context and this could have been explained.

Coastal landscapes and change

Question 2

Study Figure 3 which shows a sketch map of the area around Porlock Bay, Somerset.

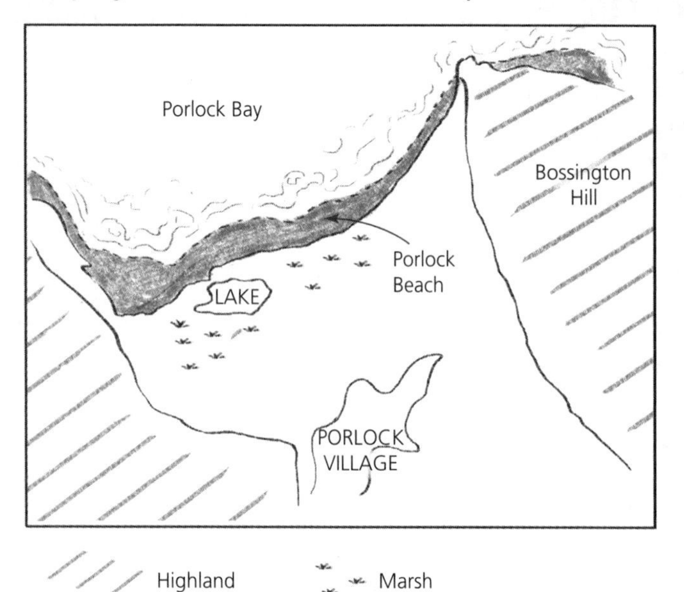

Figure 3 Sketch map of Porlock Bay, Somerset

(a) A student decided to investigate the changes in the height of the storm beach and the roundness of pebbles on that beach, in a transect from the south-west to the north-east. The prevailing wind on the beach is from the south-west. The results of the data collected are shown in the table below.

Distance from SW edge of beach (m)	Height of storm beach (m above low water mark)	Roundness index (% of particles classed as rounded)
0	4.8	18
200	7.5	20
400	10.6	35
600	9.0	16
800	10.5	22
1,000	10.0	26
1,200	13.5	44
1,400	14.0	48
1,600	13.6	70
1,800	15.0	64

(i) Using the data, calculate the mean height of the storm beach. (1 mark)

ⓔ One mark for correct calculation.

Student answer

(a) (i) The mean height of the storm beach is 10.85 m.

e **1/1 mark awarded** The student has correctly calculated the height.

The student carried out a Spearman's rank correlation test on the relationship between roundness index of the pebbles with distance along the beach.

First she stated her hypothesis: ***As distance from the south-west increases the roundness index of the pebbles increases.***

Her null hypothesis therefore was: ***There is no significant relationship between roundness of the pebbles with distance from the south-west along the beach.***

The outcomes of the test are shown in the table below:

R_s value from Spearman's rank calculation	Critical value at 95% confidence level
0.855	0.564

(ii) Which hypothesis should the student accept and why? (2 marks)

e One mark for stating the correct hypothesis. One mark for stating that the R_s value is higher than the critical value.

Student answer

(a) (ii) The correct hypothesis is that as distance from the south-west increases then the roundness of the pebbles increases. I can say this as the outcome of the test, the R_s value of 0.855, is a larger number than the critical value of 0.564. The result is therefore statistically significant.

e **2/2 marks awarded** The student identifies the correct hypothesis, and gives the right reason for this.

(iii) Suggest one geographical reason for your answer to (ii) above. (2 marks)

e One mark for identifying a reason why roundness increases and a further mark for justification of why this is the case.

Student answer

(a) (iii) The sediment is more rounded towards the north-east because they are moved in that direction, pushed along by the prevailing winds and longshore drift **a** and as this happens they are eroded by bumping into each other **b**. Their rough edges are also smoothed **b** in this process and hence they become more rounded.

e **2/2 marks awarded** The student provides a reason for difference **a**, together with two justifications **b**. The student has made good use of information given in the stem of the question.

(b) The student collected the data during a one-day field trip in summer. Explain how the student could have improved her study by collecting further data. (4 marks)

e One mark for each statement of data collection methodology, as long as the methodology is appropriate in the context of the fieldwork activity.

Student answer

(b) The student could have returned at a different time of the year to collect more data [a], for example in the following autumn when some larger storms may have had another impact on the area. This will allow the student to see how the beach system changes in different seasons [a]. The student could also visit more sample sites, and use a less systematic form of sampling — her sampling sites were every 200 m [a]. This would have increased, and thereby improved, her spatial coverage of the beach [a].

4/4 marks awarded The student makes four valid statements [a].

(c) You have also carried out field research to investigate coastal landscapes and change. Evaluate the extent to which your findings reflected the purpose of the enquiry. (9 marks)

Location of geographical investigation:

See mark scheme on p. 93.

Student answer

(c) Location of geographical investigation: The fieldwork I conducted was at Porlock Bay, a coincidence bearing in mind the other data.

The aim of my enquiry was to investigate the coastal processes and beach characteristics of Porlock Bay, Somerset. My enquiry was to find out whether the beach characteristics changed from Hurlstone Point to Gore Point as you would expect to happen if the concept of longshore drift operated at Porlock Bay. If this process was present I would expect to find larger angular boulders and beach material at Hurlstone Point where the beach is near to the cliffs that the sediment is eroded from, and smaller more rounded pebbles at Gore Point at the other end of the beach after longshore drift had transported them along shore. Therefore the purpose of my enquiry was to see whether this hypothesis was correct [a].

My findings helped to prove my hypothesis correct in that if longshore drift was in operation the beach material would be smaller and more rounded at Gore Point than at Hurlstone Point [b]. The change in average particle size between the two sites was 5.25 cm on average [b]. This is a significant change in size and helps to support my hypothesis.

My enquiry could be improved, however, by measuring from another site halfway between Hurlstone and Gore Points so that the results from the data don't just show a stark contrast. They might then show a gradual change as this is what I would expect to happen if longshore drift occurred [c]. Also the roundness of the beach material was more rounded and less angular at Gore Point than at Hurlstone Point suggesting that the particles had been transported by longshore drift and eroded in the process by abrasion and attrition, making these smaller and more rounded [d].

e **8/9 marks awarded** The student begins by providing a clear statement of the purpose, or aim, of the enquiry. It does not matter that this is the same form of investigation that appears in the early part of this question — this may occur frequently as some forms of fieldwork in a coastal context are common **a**. Another good feature of this first paragraph is that place names are stated clearly, and that there is a clear sense that the student undertook this activity. The second paragraph begins with a clear statement that addresses the evaluative aspect of the question. Furthermore, one key result is provided **b**. This is then followed by an evaluation of the investigation process (as required in the mark scheme) **c**. The answer finishes with a further statement of outcome, together with a geographical reason for it **d**. The answer addresses the question even though it is written in a somewhat simple manner.

Regenerating places

Question 3

Study Figure 4, a photograph of an area in London that has recently undergone regeneration.

Figure 4 An area of recent regeneration in London

(a) **A student decided to investigate the attitudes of a variety of people to the regeneration shown in Figure 4.**

(i) **Suggest two appropriate field questions that could be asked for this investigation.** (2 marks)

e One mark for each correct and appropriate field question.

Student answer

(a) (i) Question 1: Do you think that the environmental quality of the area has improved?

Question 2: Please will you give a reason for your answer to Question 1?

ⓔ **2/2 marks awarded** Two valid questions have been posed — one closed and one open.

(ii) The student then decided to present the data collected from the questions according to the gender, age and ethnicity of the respondee. Identify an appropriate graphical technique for displaying such data, and explain one strength and one weakness of that technique. (3 marks)

ⓔ One mark for correct identification of technique, and one mark for strength/weakness.

Student answer

(a) (ii) I would use a number of pie graphs, one for each of gender, age and ethnicity. An advantage is that you can clearly see the relative proportions of the sectors within the graph. It is visually strong. A weakness is that if there were too many categories of age or ethnicity, then the graph would become difficult to draw.

ⓔ **3/3 marks awarded** A correct technique has been identified along with a strength and a weakness.

(b) The student decided to extend her/his investigation by using social media. Give two advantages and two disadvantages of using social media for this type of work. (4 marks)

ⓔ One mark for each correct and appropriate advantage/disadvantage.

Student answer

(b) Advantage 1: Opinions can sometimes be more 'genuine' as people have taken the time out to input their opinions on to an online platform. It is something they have initiated rather than been asked about.

Advantage 2: Social media is more likely to give statements of lived experience.

Disadvantage 1: The views expressed may not be authentic, as they can be anonymous.

Disadvantage 2: The views expressed may be difficult to process for a form of analysis.

ⓔ **4/4 marks awarded** Two valid advantages/disadvantages have been provided.

(c) You have also carried out field research to investigate Regenerating places. Evaluate the success of your fieldwork experience and explain how you would make use of an opportunity to revisit the location to develop your enquiry further. (9 marks)

Location of geographical investigation:

____

e See mark scheme on p. 93.

Student answer

(c) Location of geographical investigation: The location of my fieldwork was Brockley in south-east London.

We systematically sampled ten roads in Brockley including Braxfield Road and Dalrymple Road covering a study area of 1.5 km^2. We used a scoring method consisting of a ranking based system to score every 5th house along each road out of 45. This overall score is calculated by adding up scores of 1–5 for nine different categories, judging features seen on the external view of the property that would indicate whether gentrification was occurring or not. These included categories such as: whether the original sash windows have been restored or were broken; whether there has been a loft conversion or no signs of renovation; and whether the paintwork was tidy or fading. These categories relate to the aim of our investigation because high overall scores, e.g. above 30, such as a property on Foxberry Road which scored 43 would indicate that gentrification is occurring in Brockley. On the other hand low scores on properties such as a house on Comerford Road which scored 18 would indicate that gentrification is not occurring. Based on overall scores of the properties, an average score for the road can be calculated and compared with other roads and an average score of the entire study area can be produced **a**.

This method was suitable and successful as systematic sampling reduces the effect of human interference on the results, reducing subjectivity and so giving more reliable results. This method was also successful as it allowed a range of categories to be scored allowing for simple comparison between properties, combined into data for roads, to look for evidence as to whether the process of gentrification was taking place in Brockley **b**.

However, different groups sampled different roads and so the results may still be subjective. Our method of sampling every 5th house meant that patterns of gentrification may have been missed. This could be improved on a revisit to the area by reducing the sample interval and sampling every two houses instead. For example, in our investigation we missed a house on Comerford Road, which sold for £800,000 in June 2016 on RightMove which probably (given the above-average selling price) showed aspects of gentrification **c**.

Furthermore some properties on Comerford Road were infill housing, which are relatively new compared with the Victorian/Edwardian housing targeted by gentrifiers and so did not show signs of ageing, making them hard to score using our method. On a revisit to the area, we should have identified the houses/buildings that would not fit into the category of suitable for gentrification, and therefore adopt a more stratified sampling method. In addition to this, housing on Beecroft Road had been converted into flats. This was also hard to score using our method as no individual living there is designated to take care of the external features. Therefore to provide an accurate representation of the intentions of the owners perhaps the interior of the property would have had to be viewed, or their views sampled using a resident's questionnaire. The former is probably impossible, but a revisit to the area may allow the latter to take place **d**.

> In summary, to improve this enquiry we would need to have made a more complicated or extended survey with more categories or sophisticated methods of data collection which would increase the accuracy and reliability of our results e.

e **9/9 marks awarded** Within the time constraints of the examination, this is a highly detailed and appropriate answer. The first paragraph provides a detailed account of the methodology used with some statements of outcome a. The second paragraph then links the previous one to the theme of success b. The student then moves on to the second element of the question — the opportunity to revisit an area and what could be done and why c. The following paragraph is even more detailed in its discussion of a new, or developed, enquiry, with even a sense of evaluation of new methods d. The answer is rounded off with a neat conclusion e.

Diverse places

Question 4

Study Figure 5 which shows a compound bar graph of the demographic characteristics of two boroughs of London compared with London as a whole, and England.

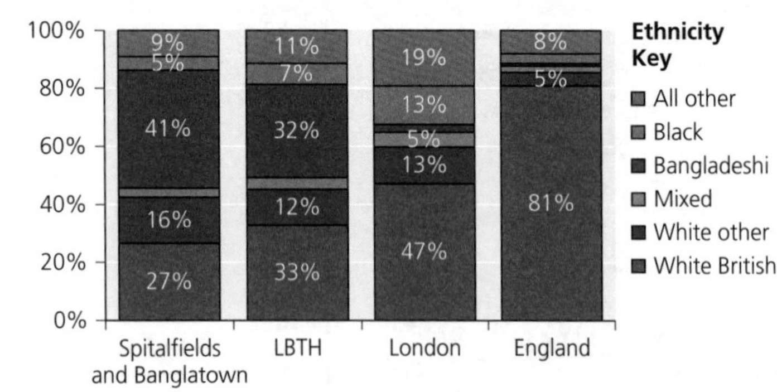

Figure 5 Ethnicity characteristics of Spitalfields and Banglatown ward compared with the London Borough of Tower Hamlets (LBTH), London, and England

(a) (i) Suggest two reasons why this technique is useful for presenting such data. (2 marks)

e One mark for each correct and appropriate reason.

> **Student answer**
>
> **(a) (i)** The divided bar graphs show data that have been converted into percentages to enable easier comparison of the proportions of different ethnicities within the places, at all scales. Once constructed, therefore, you can see clearly with the use of the same colours and the scale, the varying proportions of different ethnicities in these areas.

e **2/2 marks awarded** Two valid reasons have been given.

(ii) The student decided to carry out a Chi-squared test for Spitalfields and Banglatown, and the London Borough of Tower Hamlets, to see how the proportions of ethnicities varied from London as a whole. State the two hypotheses that she set. (2 marks)

e One mark for each of the main hypothesis and the null hypothesis.

Student answer

(a) (ii) The main hypothesis is that there is a significant difference in the distribution of different ethnic groups between Spitalfields and Banglatown/LBTH and London.

The null hypothesis is there is no significant difference in the distribution of different ethnicities between the areas.

e **2/2 marks awarded** Two correct hypotheses have been given.

(iii) Explain why the Chi-squared test is an appropriate test to use for these data. (3 marks)

e One mark for each valid reason with a further mark for expansion to a maximum of 3 marks.

Student answer

(a) (iii) This test is appropriate as the data are organised into categories which show observed values for each of the two London boroughs (O in the test) and expected (or E) values for London as a whole a. The data are also organised into frequencies in that actual numbers are not used but rather there are percentage figures for each area a. Consequently the relative differences in the data between the areas can be identified. In short, the numbers can be placed in the appropriate table, and a result calculated a.

e **3/3 marks awarded** Three valid statements are made.

(b) Identify an appropriate alternative graphical technique for displaying the data for London and England, and outline a disadvantage of this alternative technique for these data. (2 marks)

e One mark for identifying an alternative technique, and one mark for a disadvantage.

Student answer

(b) The data for London and England could have been presented by a pie chart. However, there are some very small proportions on both graphs for London and England and these would be difficult to show accurately on a pie graph.

e **2/2 marks awarded** The student provides an appropriate alternative technique, and then recognises a valid practical difficulty of using it for these data.

(c) **You have also carried out field research to investigate Diverse places. Evaluate the success of your investigation and the extent to which it has improved your geographical understanding of the area you studied.** (9 marks)

Location of geographical investigation:

---

e See mark scheme on p. 93.

Student answer

(c) Location of geographical investigation: The fieldwork I did was based in Morecambe.

The investigation meant that I found out more about social deprivation and inequalities in an area that has suffered decline and unemployment **a**. My results showed me that large areas in Morecambe have been affected by the decline of tourism in the area. Large areas were run down and houses were in poor condition as people don't have the money to maintain them. My findings also showed me that some areas were not affected, e.g. Tarrisholme, which did not suffer from social inequality.

This area had large well kept houses with large, neat gardens and the residents were wealthy showing that social deprivation does not affect all areas. The areas most affected were closer to the town centre of Morecambe in an area called West End. This has suffered most decline because the residents have become unemployed because of the closure of businesses **b**. Therefore in being able to see these diversities within the town, I can judge my enquiry to be a success.

The changing behaviour of tourists means that Morecambe no longer caters for people who stay there. It is visited by people who are there for the day. This means that several of the premises are boarded up and many of the boarding houses cater for people under the social services, who largely come from other towns and cities in the north-west of England such as Preston and Bolton **c**. The fact that people did not have a sense of ownership of the place means that it was run down and there was a lot of graffiti and rubbish on the streets. Areas that weren't affected were in the suburbs where the residents were not employed in the tourist industry and therefore were not affected and could maintain their house. I learnt that social deprivation does not affect all areas but it does mean that people's quality of life is affected when they are unemployed and live in poor environments **d**.

e 5/9 marks awarded The student begins by providing a clear, though general, statement of the outcome of the enquiry **a**. This is then followed by a series of somewhat simplistic statements of results with some generic statements of reasons behind them **b**. On the other hand, the student does identify variations, or diversity, within the town with specific references to Tarrisholme and West End. The second paragraph of the answer seeks to address the second part of the question, and here again some valid points are made as to why there are social variations within the town. The student links these to the changing nature of a main economic activity in the town, tourism, and to wider social influences such as the housing of 'social services' people from other towns **c**. The answer ends with further generic and simple statements **d**. The lack of detail in the answer, and the rather generic statements made, may suggest that the enquiry itself was too straightforward. This of course could have been down to a decision made by the supervising teacher.

Knowledge check answers

1 **Systematic:** advantages — can give a good coverage of an area, straightforward to undertake; disadvantages — can potentially miss key areas when undertaken along a line/transect, this can lead to under- or over-representation of certain groups or features.

Random: advantages — minimises bias and hence error in the sample; disadvantages — can lead to clustering and/or gaps in a total population depending on where the 'numbers' fall, time-consuming to undertake.

Stratified: advantages — reduces potential for bias in areas where a variation within a population exists; disadvantages — can be difficult to obtain information on known variations within a population so the task can be challenging.

2 The population distribution of a country: choropleth or dot map

The distribution of hospitals in a country: dot map

The variations in ethnicity in a city: choropleth map

The origins of customers to a large supermarket: desire line map (if large numbers); trip line map (if small numbers)

The number of health professionals per 1,000 people around the world: choropleth map

Travel times from a major city such as London: a form of isoline map (called an isochrone map)

The main directions of migrations around the world: flow line or desire line map

3 The changing length of a glacier/sand spit/alluvial fan over time: line graph

The percentage of global deaths from a range of non-communicable diseases: bar graph (simple or compound), or pie graph

The energy mix of a country: compound bar graph or pie graph

The top ten countries with the highest urban populations: bar or pie graphs located on a world map

The potential relationship between life expectancy and health expenditure of a nation: scatter graph

4 The IQR determines dispersion around the median, whereas the SD measures dispersion around the mean.

5 The modal values for each of the columns x_1 and x_2 are:

$x_1 = 7$; $x_2 = 14$

6 The ranges for each of the columns A and B are:

A = 13.8; B = 16.8

7 Changes in pebble sizes from one end of a beach to another: measures of central tendency; Spearman's rank correlation

Examining the differences in the distribution of different ethnic groups within wards in a city: Chi-squared test

How the concentration of PM_{10} particles changes with distance from the centre of an urban area: Spearman's rank correlation

Examining the varying orientation of the long axes of drumlins in two areas of study: Chi-squared test

Examining the potential variation in sediment sizes of fluvioglacial deposits at two locations: Student's *t*-test

Note: **bold** page numbers refer to definitions

A

accuracy **53**, 63
actions and attitudes (synoptic theme) 82, 83
aerial photographs 10
A-level exam
  fieldwork (*see also* independent investigation) 54
  synoptic skills 84–92
anomalies, data **69**
appendices 72
AS exam 5
  fieldwork 49–54
  fieldwork questions 50–51, 52, 93–107
  synoptic skills 83–84
assessment 5
  fieldwork skills 51, 93
  independent investigation 56, 61, 66, 69
  in synoptic questions 86
  synoptic skills 84
assessment objectives 51

B

bar graphs 24–25
bias
  data sources 8
  questions 14
  sampling 13
bibliographies 73
big data **64**

C

carbon cycle 80
carbon dioxide emissions 91
cartographical techniques 15–22
case studies, smart use 86
central tendency measures 32–33
Chi-squared test 44–46
choropleth maps 19–20
coastal landscapes 75, 98–101
coding 12
comparative bar graphs 24
comparative line graphs 23
compound bar graphs 24, 25
compound line graphs 23, 24
conclusions
  fieldwork 52
  independent investigation 69–71, 72
confidentiality, data collection 14
correlation 41–44
coursework *see* independent investigation
creative media 8–9
critical reflection 70–71
critical values **41**, 43–44, 45–46

D

data (*see also* qualitative data; quantitative data)
  anomalies 69
  collection 52, 60–66, 72
  familiar and unfamiliar **51**, 52
  lack of 60
  mapping 15–22
  presentation and analysis 66–69, 72
  primary and secondary 52, 60–61, 64–66
  statistical analysis 32–48, 68
data series, graphs 23–24
data sources 52, 64–65
  geospatial 14–15, 53–54
  qualitative 8–12
data stimulus questions 84
deciles 53
degrees of freedom **40**, 45–46
demographic trends 90
dependent variables **25**
desire line maps 17, 18
dispersion, measures of 34–39
dispersion diagrams 30–31
distribution maps 21
divergent bar graphs 24, 25
divergent line graphs 24
Diverse places (topic) 53–54, 77–78, 104–107
dot maps 21
drainage basins 79–80

E

energy security 80
enquiry process **50**, 51–53, 55, 56
environmental sustainability 88
ethical issues 14
evaluative skills 71, 86–87, 91–92
executive summaries 73
expected data 44

F

familiar data **51**, 52
feasibility studies 60
field notebooks 50
field sketches 11
fieldwork (*see also* independent investigation)
  assessment 5, 51
  data sources 52–54
  definition **50**
  enquiry process **50**, 51–53, 55, 56
  equipment 63
  ethical issues 14
  health and safety issues 58–59
  methodologies 61–66
  preparation for 51
  specification requirements 49–50, 54
  suggested themes 74–80
fieldwork questions, AS exam 50–51, 52, 93–107
flow line maps 17, 18
frequency distributions, pie charts 26–27
futures and uncertainties (synoptic theme) 82, 83, 90–91

G

geographical concepts, specialist 89
geographical information systems (GIS) 14–15, 53–54
Geopolitics (topic) 78–79
geospatial data 14–15, 53–54

Gini coefficient 36–39
GIS 14–15, 53–54
glacial landscapes 74–75, 93–97
globalisation 78–79, 90
global positioning systems (GPS) 14
global themes, synoptic approach 85, 89–90
global warming 90, 91
GPS 14
graphs 22–31

H
health and safety 58–59
human geography, fieldwork 76–79
hybrid data 52, 64
hypotheses **51**, 60
  data relevant to 63

I
independent investigation (*see also* fieldwork)
  aims and purpose 56–58, 59–61
  analysis and interpretation 66–69
  assessment criteria 56, 61, 66, 69
  conclusions and critical evaluation 69–71
  data collection 60–66
  health and safety issues 58–59
  requirements 54–55
  writing up 71–74
independent variables **25**
Index of Multiple Deprivation 53–54
indices 48
inequality 90
  measures of 36–39
inferential statistics 39–46
inter-quartile range 34–35
interviews 7–8, 63, 70
isoline maps 20, 21

J
judgements, criteria for 86–87

K
kite diagrams 29–30

L
landscape systems, fieldwork on 74–75
language
  evaluative 92
  in independent investigation 73
line graphs 23–24, 28
literature 8–9
literature sources 57
logarithmic graphs **28**
Lorenz curves 37–38

M
maps and map reading 11–12, 15–22, 53–54
  geospatial 14–15, 53–54
mark schemes 93
mass extinctions 90
mean 32–33
measurements, accuracy and reliability 47, 53
median 33
methodologies
  data collection 64, 65
  limitations 70–71
mode 33
movement, map representation 17–18

N
NEA *see* independent investigation
newspapers 8
normal distribution **35**
novels 8–9
null hypothesis **39**, 44, 45
number, understanding of 46–48

O
observed data 44
online data sources 14–15, 53–54, 64–65
oral accounts 8
Ordnance Survey maps 15

P
photographs 10
physical geography, fieldwork 74–75, 79–80
pie charts 26–27
pilot surveys 60
place
  perception 9–10, 76–77
  representation in creative media 8–9
players (synoptic theme) 82, 83
population trends 90
poverty, global 90
precision 63
presentation, independent investigation 73–74
primary data 52, 60, 61, 65
proof reading 73
proportional divided circles 26–27
proportional symbols, on maps 16–17

Q
qualitative data **52**, **61**
  coding 12
  data collection 7–8, 63
  sampling 12–13
  sources 8–12
  specification requirements 6
  strengths and weaknesses 65
quantitative data (*see also* graphs; maps) **52**, **61**
  geospatial 14–15, 53–54
  specification requirements 6–7
  statistical analysis 32–48
  strengths and weaknesses 65
quartiles 34

questionnaires 7, 14, 63, 70
questions
  biased 14
  types 7–8

R
radial diagrams 29
random sampling 13
range 34
referencing systems 57
reflection, on enquiry process 53
Regenerating places (topic) 53–54, 77–78, 101–104
regression lines 25
relational statistical techniques 39–46
reliability **53**, 63
report writing 71–74
representative samples 13
research questions 57–58, 59–60
residuals **26**
resource booklets, for A-level exam 84–85
resource crisis, global 90
risk assessment 58–59
rose diagrams 29
rural fieldwork 77–78

S
samples
  inaccuracy 39
  size 13, 62
sampling frames 62
sampling strategies, problems 70–71
sampling techniques 12–13, 62–63
satellite images 10
satellites, GPS 14
scatter graphs 25–26
  correlation test 41–44
secondary data 52, 60–61, 64–65, 66
Shaping places (topic), fieldwork 76–78
significance **40**, 43–44, 45
sketch maps 11–12
skills and techniques
  assessment 5
  specification requirements 6–7
social media 9–10
socio-economic data 53–54
spatial distributions, Chi-squared test 44–46
spatial patterns, maps 19–21
Spearman's rank correlation 41–44
standard deviation 35–36
standard error **39**
statistical techniques 32–48, 68
  limitations 70–71
  specification requirements 7
stratified sampling 13
Student's *t*-test 39–41
super output areas (SOAs) 53
sustainability 88
synoptic charts 15–16
synoptic skills and themes 82–92
systematic sampling 13

T
tables, data presentation in 66
teachers, role 57–58, 67, 71, 74
technological change 90
tectonic hazards, fieldwork 75
television and film 9
tertiary data 52
theories and models, synoptic approach 85
time series graphs 23–24
triangular graphs 27–28
trip line maps 17, 19

U
unfamiliar data **51**
urban fieldwork 77–78

V
validity **53**
variables, dependent and independent **25**

W
water cycle, fieldwork 79–80
weather maps 15–16
word count, independent investigation 71
writing style 73

X
x-axis **22**

Y
y-axis **22**